# Silent Earth

## Will Humans Give Up Fossil Fuels?

## James Powell

Benjamin Franklin Medalist for Engineering

## With Jesse Powell and James Jordan

Silent Earth
Will Humans Give Up Fossil Fuels?

ISBN-13:
978-1533110053

ISBN-10:
1533110050

Library of Congress Control Number: 2016908153


Printed in the United States

For information about bulk purchases, please contact:
james.jordan@magneticglide.com

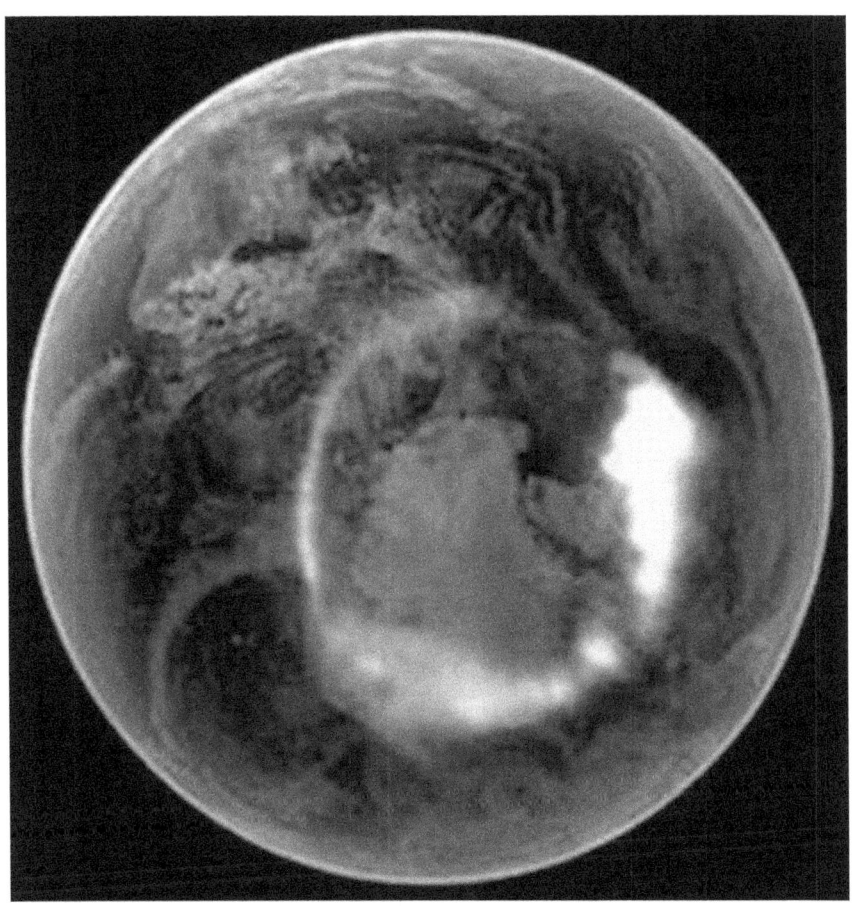

The cover image of the Aurora australis (11 September 2005) as captured by NASA's satellite, digitally overlaid onto a composite photograph of the Earth symbolizes the beautiful solution proposed in this book to the threat posed by combustion of fossil fuels to the future of civilization. The Aurora is observable proof that Earth is bathed in Energy generated by the Sun that can be converted to electricity without combustion.

## Other Books by James Powell and Gordon Danby

The Fight for Maglev, Making America the World Leader
in 21st Century Transport

Maglev America, How Maglev Will Transform the World
Economy

StarTram, The New Race to Space

# Dedication

To the Passionate Pursuit of an Idea

Rachel Carson, through her monumental book, "Silent Spring", inspired society to become aware of the environmental dangers of pesticides to birds, humans, and all of nature. Silent Spring led to the ban of DDT for agriculture uses and the formation of the Environmental Protection Agency. For all of humanity, thank you, Rachel Carson. We dedicate *Silent Earth* to you.

# The Pathway to a Non-Fossil Energy Future

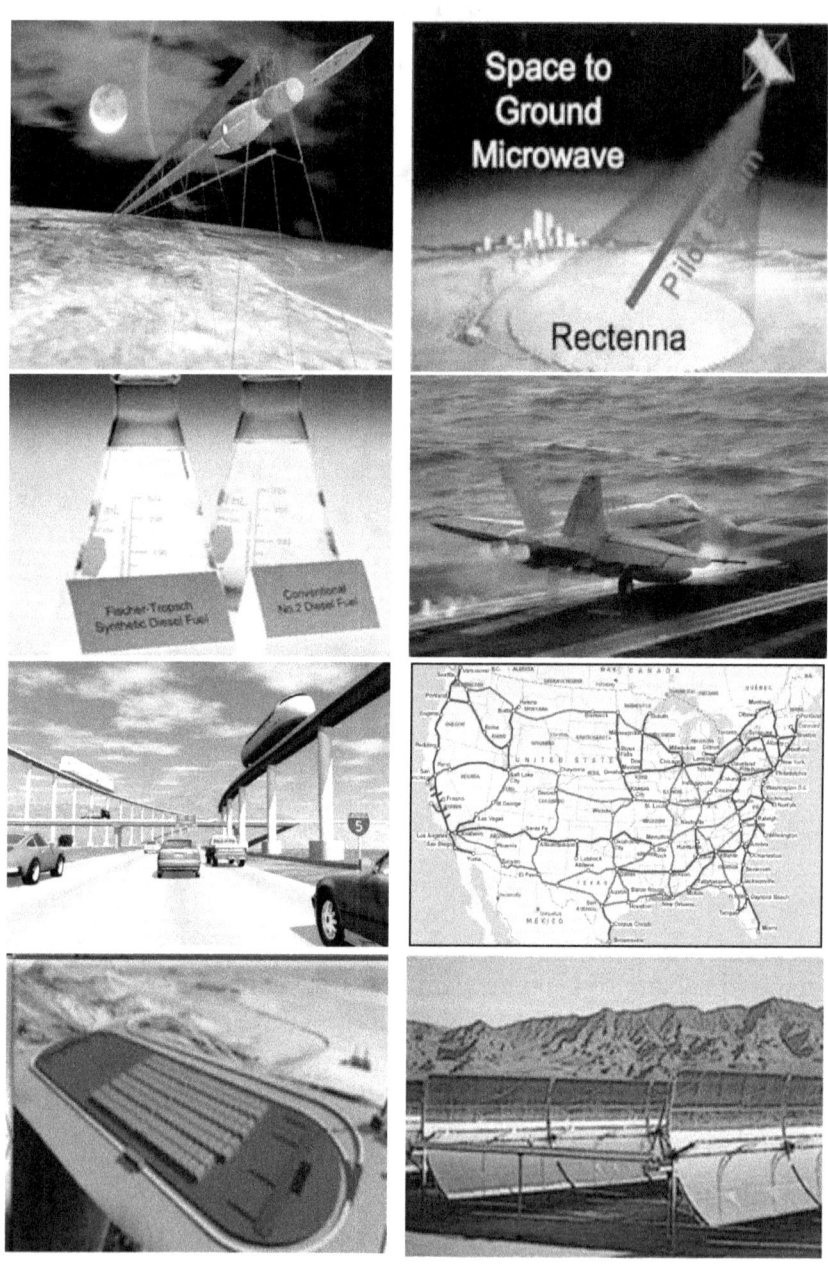

# To Our Readers

In the last 200 years, human life has undergone an incredible transformation, with capabilities and experiences undreamt of in 1800. Before then, humans lived off the land for food and energy. As we began to harvest massive amounts of energy from fossil fuels—coal, oil, and natural gas—the world radically changed.

We have seen great leaps forward in quality of life:

- For transportation, we have gone from using horses and wagons to trains, autos, airplanes, and Maglev.
- Instead of using wood fires for heating and candles for lighting, we use electric heating and lighting.
- Our food supply has become safer as we moved from unreliable, unrefrigerated, unsafe farm produce to safe, refrigerated, reliable, diverse food supplies.
- Goods are now mass-produced with the help of robots, instead of being manufactured by hand.
- To process data, we use powerful, ultra-fast computers instead of hand calculators.
- Before, we could only disseminate knowledge slowly, through printed books and newspapers. Now, we have radio, movies, TV, telephones, cell phones, and the internet.
- People have moved from an isolated life to a more fully-connected life.

Ever-increasing energy production has made phenomenal technological progress possible. So far, we have been able to use fossil fuels to power this progress. But fossil fuels are a limited resource. And if we continue using fossil fuels at the current ever-increasing rate, the result will be widespread environmental catastrophe via global warming, sea level rise, and acidification of the oceans.

Humanity is now in a situation in which we must consume massive amounts of energy to sustain society—but cannot continue using fossil fuels as the source of that energy. How do we overcome this problem?

First, we explain what *won't* work. We won't be able to conserve our way out of the problem by exhorting people to use less energy. It's simply not realistic to expect people to use less energy if it means going back to a lower standard of living. Conservation and efficiency efforts can generate some energy savings, but not anywhere near enough to meet demand.

We also won't be able to meet our energy needs using existing alternative energy sources. Even if you add up all the energy from biofuels, nuclear, wind and ground-based solar, it will not be enough to meet projected demand. Therefore, we need a new strategy. This book explores the possibility of using superconducting magnetic levitation (superconducting Maglev) technologies for transportation, power generation, and storage.

Superconducting Maglev transport, which was invented 50 years ago by James Powell and Gordon Danby at Brookhaven National Laboratory, does not need fossil fuels. Levitated and propelled by electricity, Maglev vehicles travel safely at very high speeds, without friction. They are very energy efficient and emit no pollution or greenhouse gases. Superconducting Maglev vehicles are already in use in Japan. In the years ahead, superconducting Maglev will transport passengers and freight cleanly at low cost within and between countries all around the world.

Maglev also has groundbreaking applications beyond transportation. Superconducting Maglev can launch payloads at much lower cost and in much greater volumes than rockets can. This allows for very low-cost electric power to be beamed to Earth from satellites in orbit. Maglev-based energy storage systems can then store and release this power to meet time-varying energy demands. This will greatly reduce fossil fuel consumption.

Jim Powell and Gordon Danby, the inventors of superconducting Maglev, believe that major development of Maglev technologies can significantly reduce the need for fossil fuels, helping humanity avoid serious, widespread environmental disaster.

First, let's look at the current predicament.

## Overuse of fossil fuels will cause environmental catastrophe and societal collapse.

Fuels have dramatically improved our standard of living and the human lifespan. All of us, our children and grandchildren, and their children and grandchildren, can continue on the path of an even better life here on Earth and beyond. But as we discuss in the book, the world's climate and food supply is at risk of environmental catastrophe from the release of greenhouse gases emitted by the combustion of fossil fuels. If we are to continue toward the goal of an ever-better life, we must transition from fossil fuels to new sources of sustainable energy that will not lead to environmental catastrophe.

Today, the world consumption of fossil fuels is immense, and at the present growth rates for the world economy, will almost double by 2050—only 34 years from now. World GDP will increase from 70 trillion dollars, to over 200 trillion dollars per year by 2050. Today's 1 billion automobiles in the world will be 2.5 billion in 2050. World electricity generation will more than double by 2050, along with world demand for other applications, transport, industrial, and residential and commercial uses.

Such a scenario will result in environmental catastrophe and extinction of most of Earth's life forms.

1. The acidity of Earth's oceans has already increased 30% since we began burning fossil fuels due to the absorption from emissions of carbon dioxide. Ocean acidity will reach the point where marine organisms cannot form protective shells. They—along with all of the other species that depend on them for food—will go extinct. The oceans will become lifeless deserts.

2. Runaway global warming will start and exponentially grow, even if we stop burning fossil fuels. There is more carbon stored as organic carbon compounds in the frozen Arctic permafrost and in unstable methane hydrate deposits on the sea floor, than there is carbon dioxide now in Earth's atmosphere.

As Earth temperatures increase, the permafrost melts, releasing methane and carbon dioxide. Methane is 20 times more potent as a greenhouse gas than carbon dioxide. Already, carbon dioxide and methane emissions are at dangerous levels. As the emissions from the thawing permafrost and the Arctic Ocean continue to increase, the process will "runaway" and warm the planet to catastrophic levels—even if we completely stop using fossil fuels. When will this occur? Nobody knows, but possibly well before 2100 AD.

## People won't cut back on fossil fuels if it means a lower standard of living.

It is clear that we must reduce our use of fossil fuels. But if using less energy means a lower standard of living, it won't happen.

Readers, think back to how your daily life experiences have changed since you were born.

Life was much simpler back then, from transportation to communications to food production to climate control. We used a lot less energy because our more primitive technologies didn't consume a lot of energy. But we also had a much lower standard of living.

To illustrate how much life has changed since the 1930's, we can look at the life of our oldest author, James Powell. Jim, as friends call him, was born 83 years ago in Rochester, Pennsylvania, not far from Pittsburgh. That year was the final production year for the Ford Model A automobile. The total number of US autos in 1932 was 20 million. The total number today? 200 million.

What about airplane travel? In 1932, 474,000 US passengers traveled by air. In 2015, 800 million US passengers traveled by air, more than 1500 times as many.

In those days, travel outside one's local region was rare. In Jim's small town, there was no bus service, just walking or hitching a ride in somebody's car. To get to high school, Jim usually caught a ride with a neighbor to travel the 17 miles to Pittsburgh, attended classes, then hitch-hiked home. A few times a year, Jim's family would drive or take the local train to visit extended family in nearby towns. Once in a very great while, they would travel to Pittsburgh. If you were lucky you could get great views, if the coal dust smog wasn't heavy that day.

Life in Jim's small town was not much like today. You could make phone calls on the single rotary dial telephone in the house, but because it was a party line, you had to wait until the line was clear. In small towns, radio and telephone were it for electronic communication. There were no cellphones, TV, internet or movies. For food, Jim's mother canned fruits, vegetables and meat. For heat, a coal-fired furnace heated the house by sending a big blast of hot air up through a floor vent in the center of the house.

After college, Jim went on to graduate school at MIT in 1953, becoming their first doctorate in nuclear engineering. He did his doctoral thesis on the fastest computer in the world at the time, the Whirlwind computer. It had no transistors, just 5,000 vacuum tubes in a 4 story building, with a few thousand bytes storage. Programs were written on many punched paper tape rolls. One error in punching the tape meant many hours of work was lost.

People won't willingly go back to a simpler way of life. It is critical to understand these social realities:

- Humanity will NOT give up fossil fuels if it means a lower standard of living. We will not go back to the 18th Century before we started massively consuming fossil fuels. Morcover, people in developing countries like China and India want to raise their standard of living to be like present living standards in developed countries. If it means increasing fossil fuel consumption, that's what people will do.
- Appeals to humanity's better nature — "save the environment" or "think of our grandchildren"—will not slow the increasing consumption of fossil fuels.
- The only way to stop massive consumption of fossil fuels for electricity generation, transport, industries, residential and commercial uses, is to develop and implement new, and

practical non-fossil fuel energy sources that are more economical, both in investment and operation, than fossil fuels. Otherwise, people and nations will continue to burn fossil fuels.

## Existing alternative energies are not enough to meet increasing demand.

*Silent Earth* examines what options humanity has to transition from fossil fuel energy sources to clean, renewable energy sources that can meet our future energy needs and avoid environmental catastrophe. What new energy sources can replace fossil fuels?

First, as discussed, humanity will not give up fossil fuels unless the new energy sources can maintain or improve our present standard of living and have favorable economics with net savings for the public. The social and environmental effects of introducing new energy sources must be favorable and not disruptive to society as a whole, though there may be some negative effects to certain segments of society.

Second, the new source must be capable of supplying the vast quantities of energy that the world will need in the coming decades. Consider just the electric energy we will need by 2050—just 34 years from now.

In 2014, the world generated 22,700 terawatt hours of electricity. One Terawatt Hour equals 1 billion kilowatt hours. World electrical generation is projected to grow to about 54,000 Terawatt hours per year by 2050 AD. That's 6,000 Kilowatt hours per year per capita for the 9 billion people on Earth.

To supply the 54,000 Terawatt hours of electrical generation, what are the existing options?

- Biofuels? These do not generate enough energy. Even all the arable land in the world would not be enough. And arable land will need to be cultivated to feed the growing global population.
- Nuclear reactors? This would require 6,500 reactors, each 1,000 megawatts. Currently, there are only 437 reactors in the world. Limits on uranium resources, safety, and waste disposal

problems make nuclear reactors impractical on the scale required.

- Wind? A 3 megawatt peak capacity large wind turbine will only generate an average power of about 1 megawatt, because the wind does not blow all the time. At 1 megawatt average power, to generate 54,000 terawatt hours in 2050 AD, the world would need 6.5 million wind turbines. Each turbine would be 200 feet higher than the Statue of Liberty's torch.
- Ground-based solar power plants? Like wind turbines, the average power output from a large ground solar power plant is much less than the peak power it produces—only about 1/4$^{th}$—because the sun doesn't always shine, and there are clouds and storms. To generate a world total of 54,000 terawatt hours of electrical power, annually, would require on the order of 500,000 square miles, or 1/6$^{th}$ of the total land area of the lower 48 states in the continental US.

To meet energy demand while avoiding environmental disaster, we cannot rely solely on existing alternative energy technologies. We need another approach.

## Maglev technologies can power the future.

Superconducting Maglev technologies can drastically reduce the use of fossil fuels by:

(1) providing energy-efficient transportation, accounting for a big portion of fossil fuel use;

(2) enabling the low-cost launch of solar electrical generators into orbit, which can generate solar energy to be beamed back to Earth;

(3) efficiently storing the solar power, which can then be distributed to meet power demands and create synthetic fuel.

Maglev is an old idea, proposed by Emile Bachelet in France in the early 1900's, using induction from AC current in loops to levitate vehicles above a conducting surface. Robert Goddard, the father of modern rockets, also proposed Maglev.

What makes Maglev practical and very important, however, is superconductivity.

Superconductors can carry very large electric currents with zero resistance losses, indefinitely, if refrigerated at low temperatures. Using superconductors, very powerful, lossless magnets can be fabricated that can interact with a guideway to levitate and propel heavy vehicles at high speeds with energy efficiency.

How and when did the idea of superconducting Maglev originate? In 1959, Jim worked in the Nuclear Engineering Department at Brookhaven National Laboratory. He was working on using superconducting magnets, a very new technology, as part of a Magneto Hydro Dynamic (MHD) project to generate electric power from High Temperature Gas Cooled Reactors (HTGR).

One day, on a weekend drive up to Boston, Jim was stuck in a 5-hour traffic jam on the Throgs Neck Bridge in New York. Sitting for 5 hours in his stationary car, Jim thought there must be a better, faster way to get to Boston. Then a thought occurred to him. Why not use superconducting magnets to levitate and propel high speed trains?

Jim and Gordon Danby, who also worked with superconducting magnets in the Physics Department for new particle accelerators at Brookhaven, worked together to develop their first generation Superconducting Maglev System. Their superconducting Maglev paper at the 1966 American Society of Mechanical Engineering Conference in New York City created enormous interest. They were visited repeatedly by teams of scientists and engineers from Japan, Germany, and other countries.

Japan proceeded to develop Jim and Gordon's first generation inventions into Japan Rail's current superconducting Maglev system, for passenger transport. It is a great achievement, with speeds up to 370 mph, and over 100,000 accumulated passenger trips. Currently operating on the 27-mile test track at Yamanashi, Japan Rail plans to extend it to become a 300-mile route between

Tokyo and Osaka, to carry 100,000 passengers daily with a trip time of 1 hour.

Since then, Jim and Gordon have continued their work to help develop superconducting Maglev into an even more capable technology with multiple applications. Maglev technology can transport freight as well as passengers. It can also launch payloads into space at much lower cost than rockets, which will allow for large amounts of very low cost electric power to be beamed down to Earth from solar satellites in orbit. This energy can then be stored in Maglev energy storage systems at very low cost.

The path described in *Silent Earth* bypasses the constraints on the above non-fossil energy technologies the technical solutions and describes three non-fossil energy technologies that can greatly reduce the world's dependence on fossil fuels and help to avoid environmental catastrophe. They are technically practical, economically attractive, and environmentally desirable and can be quickly developed and implemented.

This book describes these new Maglev inventions and how they will benefit humanity. Specifically, the book describes a feasible, realistic plan to:

- Use Maglev to launch solar electric generators to geosynchronous orbit to beam 24/7 very low-cost electric power down to world-wide receivers on Earth.
- Use the very cheap electricity from space solar to make synthetic gasoline, diesel, and jet fuel from air and water to distribute through the regular fossil fuel distribution network.
- Create Superconducting Maglev transport networks that complement our existing highways and railways to provide rapid, low cost movement of freight, commuters, and passengers.

These three technology applications should be tested and demonstrated as soon as possible. The global urgency has arrived and there should not be further delay. The build-out of the 3 applications at large scale can be implemented by governments, private investors, or international cooperative institutions such as the World Bank or the United Nations or a new monetary fund to rapidly create the new manufacturing, transport services and synthetic fuel production and distribution to keep pace with World economic growth and provide good paying jobs for people currently employed in the fossil fuel industry.

James Jordan is the managing editor of *Silent Earth* and can be reached at james.jordan@magneticglide.com

# Table of Contents

DEDICATION ................................................................. V

TO OUR READERS ....................................................... VII

PREFACE: HUMANITY AT THE CROSS ROADS ...................................... XIX

PROLOGUE:  OUR FAUSTIAN BARGAIN WITH FOSSIL FUELS................. XXI

## PART I
### THE COMING CATASTROPHE

CHAPTER 1:  LOOKING BACK – LIFE BEFORE FOSSIL FUELS,
COMPARED TO LIFE TODAY................................................3

CHAPTER 2:  FOSSIL FUELS AND GLOBAL WARMING – THE TRUTHS ......19

CHAPTER 3: FOSSIL FUELS AND GLOBAL WARMING
– THE CONSEQUENCE................................................35

## PART II
### HOW MAGLEV CAN HELP STOP GLOBAL WARMING

CHAPTER 4:  MAGLEV TRANSPORT FOR AMERICA AND THE WORLD.....73

CHAPTER 5:  MAGLEV LAUNCH TO SPACE FOR POWER
BEAMING BACK TO EARTH ................................................147

CHAPTER 6: MAGLEV STORAGE OF ENERGY FROM WIND
AND GROUND SOLAR SOURCES ........................................179

CHAPTER 7: CLEAN SYNTHETIC, NON-FOSSIL FUELS FROM
AIR & WATER ................................................207

## PART III
### IMPLEMENTING MAGLEV TO STOP GLOBAL WARMING

CHAPTER 8: THE ECONOMICS OF IMPLEMENTING MAGLEV ...............239

CHAPTER 9: THE POLITICS OF IMPLEMENTING MAGLEV ......................251

ACKNOWLEDGEMENTS........................................................261

APPENDIX................................................................265

## *Second Coming*

*Turning and turning in the widening gyre*
*The falcon cannot hear the falconer;*
*Things fall apart; the centre cannot hold;*
*Mere anarchy is loosed upon the world,*
*The blood-dimmed tide is loosed, and everywhere*
*The ceremony of innocence is drowned;*
*The best lack all conviction, while the worst*
*Are full of passionate intensity.*

*Surely some revelation is at hand;*
*The Second Coming! Hardly are those words out*
*When a vast image out of Spiritus Mundi*
*Troubles my sight: somewhere in sands of the desert*
*A shape with lion body and the head of a man,*
*A gaze blank and pitiless as the sun,*
*Is moving its slow thighs, while all about it*
*Reel shadows of the indignant desert birds.*
*The darkness drops again; but now I know*
*That twenty centuries of stony sleep*
*Were vexed to nightmare by a rocking cradle,*
*And what rough beast, its hour come round at last,*
*Slouches towards Bethlehem to be born?*

*William Butler Yeats*

# PREFACE
# Humanity at the Cross Roads

Will Humans give up fossil fuels? Fossil fuels have made modern life possible. Before we began burning coal, oil, and natural gas, World population was less than a billion people. Today it is 7 Billion, soon to be 9 Billion by 2050 AD. Before fossil fuels, life was hard. Average life expectancy only 30 years, average annual income per capita less than 1/10th of today's income, hard labor for virtually everybody. You were born in one place and lived there till you died. No real travel. Maybe you read a few books or newspapers during your lifetime. No computers, internet, cell phones, TV, movies. Maybe you sent and got a few letters. Simple staple foods – wheat, potatoes, rice, fruit and vegetables. No supermarkets. Simple, small, crowded houses and rooms. Only candles and fireplaces for light, only fireplaces for heat.

Are any of us willing to go back to when we didn't have fossil fuels? Of course not. The energy from fossil fuels sustains our life style. The average American consumes 90 times as much energy from fossil fuels than the energy that he or she gets from the food they eat. It's like having 90 servants to provide you with a good life.

If consuming fossil fuels is the only way we can maintain our present life style, will we give up fossil fuels, even though they will cause environmental catastrophc and a massive Sixth Extinction to a Silent Earth in decades from now? Of course not. "Eat, drink, and be merry today, for some other person will die in the future, after you've gone." As the French aristocracy said, "Après moi, Deluge".

Today, many deny that the continued consumption of fossil fuels will lead to a Silent Earth, because they can't handle the truth. Others realize the truth, but don't care, because they won't be around when it happens.

It is very clear that humanity will not give up fossil fuels, unless there is an economically practical path toward developing non-fossil energy sources that are low enough in cost and of sufficient quantity to sustain our present life style, and can be implemented quickly. Unfortunately, the proposed sources, so far, do not appear adequate.

This book describes how Maglev, the first new mode of transport since the airplane, provides a near-term practical path towards greatly reducing humanity's dependence on fossil fuels, while at the same time achieving much better and safer electric transport, enabling massive amounts of amounts of clean, safe electric power beamed from space solar satellites, which can provide energy for the synthesis of low-cost, clean, synthetic hydrocarbon fuels and materials from natural air and water. Finally, fossil fuels are not humanity's enemy. Without them, we still would remain in a much harder World, with a much lower standard of living and much less knowledge of the vast and wonderful universe around us. We will be our own enemy if we are too stupid to move on beyond fossil fuels to a better, more wonderful World.

# Prologue

## Our Faustian Bargain with Fossil Fuels

The legend of Faust captures human willingness to accept long-term disaster in return for short term success and power. Mephistopheles carries out Faust's wishes for 24 years, followed by eternity in Hell. Faust realized he would spend eternity in Hell after 24 glorious years. Humanity, however in its Faustian Bargain with Fossil Fuels has just begun to realize that fossil fuel benefits will not last forever, and that Environmental Hell will soon be here.

Figure 1 shows the dramatic increase in World population since 1800 due to fossil fuels. Table 1.0 shows a summary of the consequences in the $CO_2$ greenhouse gas emissions, increased GDP (Gross Domestic Product) per capita, increased life expectancy, and increased population density. World population is projected to reach 9 billion by 2050 AD, with an average density of 160 people per square mile on all of Earth's land surface, including deserts, forests, arctic tundra, etc. The average distance between persons will then be only 400 feet – a bit less than 1 person per acre.

Part I of the book describes in more detail the benefits and the adverse impacts that fossil fuels have had for humanity. Chapter 1 compares how we lived before fossil fuels with how we live today, with much better transport, food, communications, etc. Chapter 2 describes the scale of our fossil fuel consumption and how it affects global warming. Chapter 3 describes the negative impacts of global warming – rising sea levels, storms, droughts, extinction of species, etc. Parts 2 and 3 of the book describes how Maglev, a completely new mode of transport, can greatly reduce humanity's need for fossil fuels, while maintaining and even increasing, humanity's present quality of life.

Humanity no longer lives in the "natural" world. It now lives in the "human" world, dominating all other life on Earth. Humanity will not return to a world without our present lifestyle. Humanity will continue to use fossil fuels, even though Environmental Hell awaits, unless new technologies, which are not dependent on fossil fuels, are developed that have the capability to maintain and improve our quality of life.

Humanity must develop and implement these new, non-fossil fuel technologies very quickly. Faust had 24 years before being dragged off to Hell. By 2040, 24 decades will have passed since 1800. As Earth warms, the tremendous amounts of carbon stored in cold permafrost regions as organic compounds, and in sea beds as marginally stable methane hydrates, are already starting to be released into the atmosphere as carbon dioxide and methane, further accelerating the global warming process beyond what humans are doing by burning fossil fuels.

At Trigger Point, the global warming process will run away – global temperatures will increase uncontrollably from the release of the stored carbon in the permafrost and the oceans, even if humans completely stopped using fossil fuels.

Have we already passed the Trigger Point? Maybe, maybe not. What is clear, however, is that we only have a short time before we reach it, and then go to Environmental Hell and human extinction. Will the trigger point be 2040, 24 decades from 1800? Nobody knows.

The rapid increases in population, annual Gross Domestic Product (GDP) per capita, as measured in Purchasing Power Parity (PPP) by constant 1990 dollars, and fossil fuel consumption, as measured by Billions of Metric tonnes of $CO_2$ combustion product emitted to the atmosphere, began in the 19th Century (1800-1900) and accelerated in the 20th Century (1900-2000). From 1900 to 2000 AD, World population and GDP (PPP) per capita increased by a factor of 4, while $CO_2$ emissions increased by a factor of 10. Life expectancy doubled from 32 years to 65, and population density, measured by persons per square mile of land surface area on Earth, including desolate desert regions, Siberia, Greenland, Antarctica, etc. increased four-fold. World population is projected to increase to 9 billion by 2050 AD.

Figure 1 is a graph of the global human population from 10,000 BC to 2000 AD, from the US Census Bureau. The graph shows the extremely rapid growth in the world population that has taken place since the 18th century. (3)

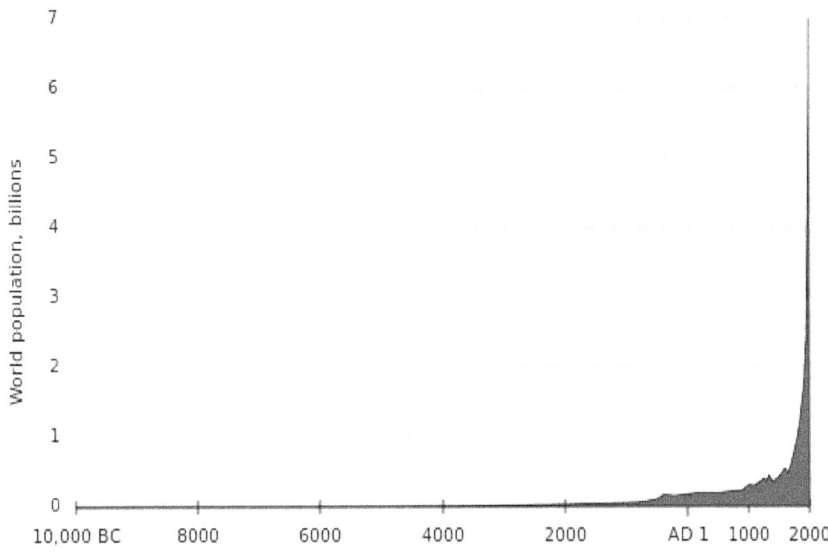

Figure 1

Table 1.0
Historical World Population (3), Fossil Fuel Consumption (4), GDP(PPP) Per Capita(2), Life Expectancy(1), Population Density

| Parameter | Year | | | | | | | |
|---|---|---|---|---|---|---|---|---|
| | 1000 | 1750 | 1800 | 1850 | 1900 | 1950 | 1975 | 2000 |
| World Population, Billions | 0.26 | 0.79 | 0.98 | 1.3 | 1.6 | 2.5 | 4.1 | 6.1 |
| $CO_2$ Emissions from Fossil Fuels, Billion Tons/year | | | 0.4 | 1.1 | 2.2 | 5.6 | 23.0 | 26.0 |
| GDP (PPP) Per Capita per Year (1900 Dollars) | $440 | $630 | $650 | $800 | $1300 | $2100 | $4200 | $6000 |
| Average Life Expectancy, years | — | 30 | 30 | 30 | 32 | 48 | 60 | 65 |
| Population Density, Persons/mile$^2$ | 5 | 14 | 17 | 23 | 28 | 44 | 72 | 107 |

# Part I

## *The Coming Catastrophe*

*Yes, these are the dog days, Fortunatus:*
*The heather lies limp and dead*
*On the mountain, the baltering torrent*
*Shrunk to a soodling thread;*
*Rusty the spears of the legion, unshaven its captain,*
*Vacant the scholar's brain*
*Under his great hat,*
*Drug though She may, the Sybil utters*
*A gush of table-chat.*
*- - - - - - - - - - - -*
*How will you look and what will you do when the basalt*
*Tombs of the sorcerers shatter*
*and their guardian megalopods*
*Come after you pitter-patter?*

*Under Sirius, W. H. Auden*

If homo sapiens is going to survive, it not only needs to learn the "Don't Sell Your Soul" lesson from Faust, but also the lesson, "Don't Be Arrogant – Pride Goes Before a Fall" from Lucifer, who was cast down from Heaven to Hell when he tried to seize power from God. Humanity needs to understand that Nature rules the World, not Humans. Figure 2 shows Dore's illustration of Lucifer's fall in John Milton's "Paradise Lost". It is small comfort when in Hell to proclaim as Lucifer did, "Better to reign in Hell, than serve. in Heaven."

# Chapter 1

## Looking Back – Life before Fossil Fuels, Compared to Life Today

Everyday life has changed incredibly since we started to consume fossil fuels. Let's take a look at life before 1800 without fossil fuels and compare it to life today with fossil fuels. When we get done comparing yesterday and today, we pose the following questions to our reader:

Faced with the choice of a) giving up fossil fuels and living as people did before we had them, or b) continuing to use fossil fuels even though they are environmentally destructive so as to maintain present life style, which choice do you think humanity will make, a) or b)?

How important are future generations to humanity today? If our children and grandchildren are going to experience environmental catastrophe from fossil fuels, will that make humanity give up fossil fuels?

If new technologies can phase out fossil fuels and still maintain and improve present life styles and not lead to environmental catastrophe, would they be supported? How quickly should they be implemented?

The seven areas we will now compare life in yesterday and today are the following:

1. Living Standard, as measured by GDP per capita
2. Transportation and Travel
3. Communication and Knowledge
4. Life Expectancy
5. Food
6. Housing and Buildings
7. Heat and Light

If the world remains committed to fossil fuels, less developed nations will not raise their living standard unless they consume more fossil fuels. Since the desire for higher living standards will prevail, future world fossil fuel consumption will greatly increase, not decrease. Efficiency increases will not be sufficient to stop the inevitable increase.

Figure 1.1 illustrates how national living standards as measured by GDP per capita, the value of goods produced annually by the country per inhabitant (1) correlates with the fossil fuel energy consumed annually per capita (2). The values shown are for 2013. GDP is expressed in US Dollars. Energy consumption is expressed in Kilograms of oil equivalent, and is the sum of the primary energy generated from the combustion of coal, oil, and natural gas.

Three important conclusions can be drawn from Figure 1.1.

- Living standard increases with fossil fuel consumption per capita. Roughly, doubling fossil fuel consumption per capita doubles GDP per capita. Developed countries with high living standards use lots of fossil fuels, on the order of 4 or more tonnes of oil equivalent (ToE) per capita per year.(1 ToE=1,000 Kg OE). Less developed countries with lower living standards use much less, India uses less than 1 ToE per year, China uses 2 ToE annually, and America uses 6.9 ToE per capita per year. To put this in perspective the primary energy input from fossil fuels to each American is equivalent to 90 times the energy he or she gets from food. It's like having 90 servants working for you.
- To achieve a higher living standard, a country must increase its energy consumption, which primarily comes from fossil fuels.
- World average Kilogram oil equivalent (KgOE) per capita is 2,000 (KgOE) annually. With a World average of GDP per capita of $10,000 per year. For a world average standard of living equal to the US standard of living in terms of annual GDP per capita, total world average annual fossil fuel consumption

would more than triple, from 2,000 KgOE per capita to 6,900 KgOE per capita.

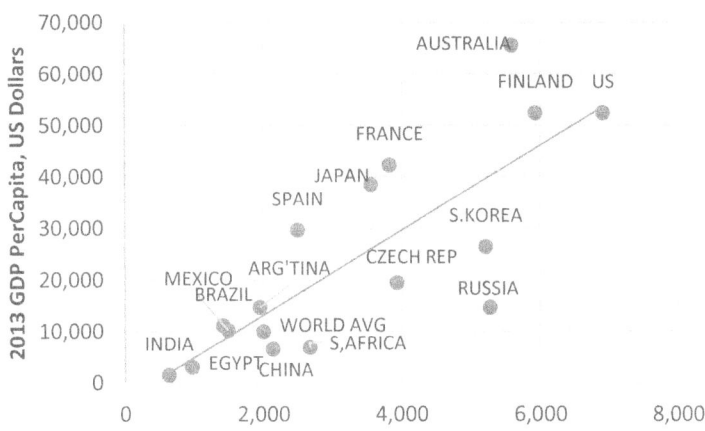

2013 Annual Primary Energy Consumption Per Capita in Kg of Oil Equivalent (KgOE)

| Country | GDP | KgOE | Country | GDP | KgOE |
|---|---|---|---|---|---|
| Australia | 65,600 | 5,592 | Russia | 14,680 | 5,283 |
| US | 52,392 | 6,909 | Brazil | 11,199 | 1,418 |
| Finland | 52,392 | 5,933 | Mexico | 10,293 | 1,492 |
| France | 42,339 | 3,827 | World Average | 10,000 | 2,000 |
| Japan | 38,528 | 3,560 | S. Africa | 6,936 | 2,675 |
| Spain | 29,685 | 2,500 | China | 6,626 | 2,143 |
| S. Korea | 26,482 | 5,222 | Egypt | 3,110 | 969 |
| Czech Rep. | 19,510 | 3,935 | India | 1,548 | 637 |
| Argentina | 14,760 | 1,953 | | | |

Figure 1.1   Gross Domestic Product Per Capita as Function of Energy Use Per Capita for Countries

5

## Transport before Fossil Fuels

Figure 1.2

Before fossil fuels and the invention of the internal combustion engine and electricity, the four modes of transport were walking, horseback, horse and oxen drawn coaches and wagons, and sailing ships. Most people walked – few had money for horses and carriages, or a trip on a sailing ship. Most people never traveled further than a few miles from home. Walking, took one hour to go 3 miles. If you rode a horse at a gallop, you could go 16 miles in one hour but taking into account average horse power, not peak horsepower, average speed was more like 6 to 8 mph.

Hauling coaches, average speed was even less, about 4 mph. With special mail coaches and good roads, in the 1790's in Britain, average coach speeds of about 10 mph were achieved. A letter from London could reach Edinburgh in 43 hours. However, most trips took longer. In colonial time, from New York City, took 4 days to get to Boston, 1 week to western Pennsylvania, and 6 weeks to Chicago. Most sailing ships averaged about 5 to 6 mph. Clipper ships weren't much faster. The Flying Cloud record trip between New York and San Francisco took 89 days, at an average speed of 9 mph.

## Transport Using Fossil Fuels

Figure 1.3

Today, we travel much further and faster than we did before fossil fuels. Internal combustion engines power autos, airplanes, trains, and ships. As technology evolves, autos and trains are transitioning to electric power instead of fossil fuel powered engines. The average American travels about 15,000 miles per year, a distance 10 or more times greater than colonial travel.

We travel 60 mph by auto, 500 mph by airplane, 200 mph by High Speed Trains, and 25 mph by ship. During his/her lifetime, the average American travels 1.2 million miles, equivalent to 50 trips around the world, and 2 round trips to the Moon.

Not only do the average Americans travel enormous distances, but we also have tremendous amounts of freight transported – fuels, food, equipment, etc. – to sustain our life style. Over an 80 year lifetime, the average American has 800,000 ton miles of freight transported in the US by truck and rail. For an average transport distance of 400 miles, that's 2,000 tons of freight to keep us fed, healthy, housed, entertained, working, and able to travel.

# Communications and Knowledge before Fossil Fuels

Figure 1.4

Before fossil fuels, people communicated through letters, newspapers, and books, sent from one person to another by whatever slow mode of transport – stagecoaches, horses, and sailing ships. In colonial times, it took 5 days to get a letter or newspaper or books from Boston to New York. By clipper ship from New York to San Francisco, it took 3 to 4 months. From London to New York, a month.

However, when Pony Express started operation in 1860, sending letters, newspapers, and books between New York City and San Francisco became much faster. Only 10 days, not 3 months to send a message, 10 more days to get a reply, for a total of 20 days, not 6 months. Pony Express stations were about 10 miles apart along the route, the distance a horse could gallop before it became tired. Riders could not weigh more than 125 pounds, and carried 20 pounds of mail. Riders changed after riding 75 to 100 miles. The mail went on day and night. The Pony Express only operated for 2 years, being supplanted by the much faster telegraph.

## Modes of Communications Today

Figure 1.5

Today, there are over 7 billion mobile phones and 3 billion Internet users in the world, with the capability to instantly send e-mail messages and phone calls anywhere in the World. No 10 days, New York City to San Francisco. Via the Internet one can obtain detailed information on current events and any subject – no need to buy a newspaper at the newsstand or go to the library. Or you watch the news on television and listen to discussions and educational programs. Or take on-line courses on your computer. Or view movies, documentaries, comedies, historical, futuristic, fantasy, science fiction, etc. *The Great Train Robbery, shown here,* is a 1903 American silent short Western film written, produced, and directed by Edwin S. Porter. Film historians now largely consider *The Great Train Robbery* to be the first American action film.

The knowledge, and the rate of flow of knowledge to today's world's population is vastly greater today than it was before fossil fuels made modern communications possible. Sometimes, one wonders if we are able to really absorb it.

## Life Expectancy before Fossil Fuels

Figure 1.6 gives the timeline, shown below, for life expectancy at birth as a function of time since the Paleolithic Age, which ended about 10,000 years ago. Humans lived as hunter-gatherer groups, with stone tools. In the Neolithic Age, humans started agriculture, domesticated cattle, chickens, pigs and pottery. The Neolithic age ended about 4,000 BC, and we transitioned to the Bronze/Iron Age, within current Egypt and Syria. Then came the classical Greek age with all the great Athenians from 500 BC to 300 BC.

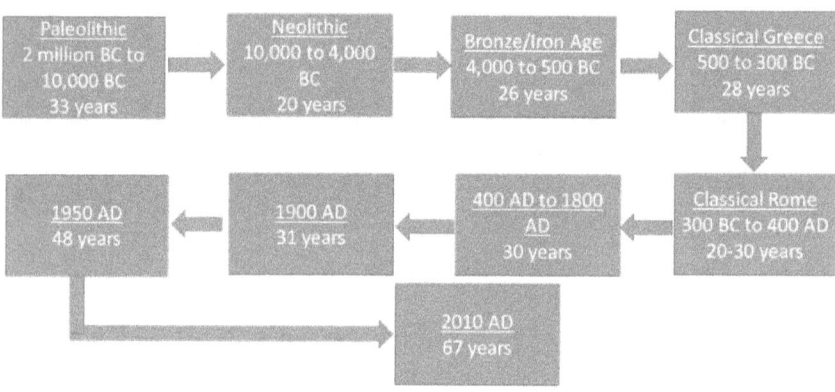

Figure 1.6 Historical Life Expectancy

Then came classical Rome, which lasted to about 400 AD. After Rome came lots of empires and societies all over the World, still with a life expectancy of about 30 years at birth for 1500 more years. During the period from ancient times to the beginning of the Industrial Revolution in the 1800's, life expectancy at birth was limited by the early deaths of young children. If you made it to your teens, you probably could live to be 60 or 70. Global life expectancy at birth, in 1800 was about 30 years. It rose slowly during the 1800's. By 1900, however, it started to take off as the improvements in health care enabled by the increased income from fossil fuel use kicked in. By 1950, life expectancy increased to 48 years, reaching 67 years today.

# Life Expectancy with Fossil Fuels

Life expectancy at birth has increased from about 30 years in the pre-fossil fuel era to over 80 years in the world's developed countries, whether in Asia, e.g. Japan, Europe, e.g. Germany, Norway, Sweden, France, etc., North America, e.g., US and Canada, and South America, e.g. Chile. (3) The poorer countries in Africa have lagged behind in life expectancy (Figure 1.7), 46 years, with an annual per capita GDP of $809. As per capita GDP increases, so does life expectancy, leveling out to a range of about 70 to 84 years for incomes greater than $10,000 per capita. As noted earlier, the increase in life expectancy from the approximately 30 years before the advent of fossil fuels to today's levels of 70 to 80 is primarily due to the much lower death rate for young children. Once countries with incomes below about $10,000 per capita achieve higher incomes, the resulting improved health care well enable them to greatly reduce children deaths, bring their average life expectancy to the 70 to 80-year range for today's developed countries.

| Country | Life Expectancy at Birth (years) | 2013 GDP/Capita (US Dollars) |
|---|---|---|
| US | 79 | 52,392 |
| Germany | 81 | 45,091 |
| Japan | 84 | 38,528 |
| Spain | 82 | 29,685 |
| Czech Republic | 78 | 19,510 |
| Chile | 80 | 15,723 |
| Argentina | 76 | 14,760 |
| Russia | 69 | 14,680 |
| China | 75 | 6,626 |
| Egypt | 71 | 3,110 |
| Indonesia | 71 | 883 |
| India | 66 | 1,548 |
| Nigeria | 54 | 2,966 |
| Angola | 51 | 5,668 |
| Haiti | 62 | 745 |
| Sierra Leone | 46 | 809 |

Figure 1.7

## Food Supplies before Fossil Fuels

Figure 1.8

Before efficient long distance transport was made possible by fossil fuels, virtually all food supplies were produced locally, consumed locally, and stored locally, without refrigeration. If the weather was good and the crop bountiful, you ate well and survived. If the weather was bad or disease and pests ruined the crop, you starved and died. There are lots of famines in human history. Remember the great Irish Potato Famine? The food staples were pretty basic.

For meat, chickens, pigs, cattle, and sheep. For grains, wheat, maize, and rice. Various fruits and vegetables, depending on where you lived. Grains, fruits, and vegetables were seasonal and had to be stored for long times if you wanted them after harvesting them – no refrigerators – hoping that the rot and rats didn't get at them. Meat animals were better – they could slaughter them when hungry. Meals were cooked at home over a wood fire or a wood stove.

## Food Supplies Using Fossil Fuels

Figure 1.9

It's a much better Food World today, enabled by fossil fuels. We can buy a much wider variety of foods in supermarkets than we could grow locally, before we had efficient long-distance fossil-fueled transport to bring us food from distant sources. We can grow food much more efficiently and cheaply using fossil-fueled farm equipment. We all don't have to be farmers – we're able to do many more things. We can freeze food and store it in electrically powered refrigerators without it spoiling. We can keep food for very long periods in metal cans and glass and plastic bottles made using fossil fuels. No worries about rot and rats. We can minimize the effort and time to prepare meals – frozen TV dinners, canned food, fast food and fancier restaurants when we're in the mood. All made possible by fossil fuels.

## Housing before Fossil Fuels

Fig. 1.10

Before fossil fuels almost all houses were all constructed using wood and stone. There was very little brick and metal for building houses. Out in the prairies, there was no wood or stone, so they used slabs of thick, strong grass sod. The sod houses were well insulated, but very damp and subject to rain damage. The outer walls were protected by stucco and wood panels, and the inner walls could be plastered. Not a type of house one would choose today. If wood were available we could build log cabin homes or fancier houses like the Schenke Grooke house built in Brooklyn in 1675 (see photo). If one were really rich, like George Washington, you could live in a mansion like Mt. Vernon (See 1796 lithograph), or the Philadelphia brick building where he served as President from 1792 to 1797 (see lithograph).

## Housing with Fossil Fuels

Figure 1.11

The type of house we live in today depends very much on 2 factors: 1) income level and 2) location. About half of Americans spend at least 30 percent of their income on housing, either as owners or renters. They can't afford to spend much more without sacrificing their quality of life – food, health care, communications, etc. In densely populated regions, they live in apartments, in suburban and low population density regions, middle income people tend to live in individual houses on individual plots of land, with individual designs. The World population is rapidly moving into cities, so most of humanity will soon live in cramped apartments.

If one is rich, one can live in large, plush, fancy houses in cities, or outside cities. The Vanderbilt's house in Manhattan an early example of mansion style houses is very large. More recent examples can be found in the Hamptons on Long Island. Large penthouses in Manhattan go for as much as 100 million dollars. If you are at the opposite end of the income ladder, the only choice may be a mobile house in a trailer park, like the example shown in the photo.

## Heat and Light before Fossil Fuels

Figure 1.12

Before fossil fuels, there weren't many chances for heat and light at night. If you were cold and had a fireplace with wood available, you could build a fire. Not very effective heating, but if you stayed close to the fire you could be warm, plus actually seeing the room around you – sort of – at night. That's with an open fireplace. With the more efficient stove invented by Benjamin Franklin, you got heat but not light. To keep heat going all night, would have to wake up every so often put more logs unto the fire, or more wood into the stove. For light, if you didn't want light in another part of the house, you would use a candle or lamp – the candle made of wax and the lamp fueled with vegetable or mineral oil. Candles, lamps, and fireplaces/stoves made life better for humans for thousands of years, but they are pretty limited in their capability.

## Heat and Light Using Fossil Fuels

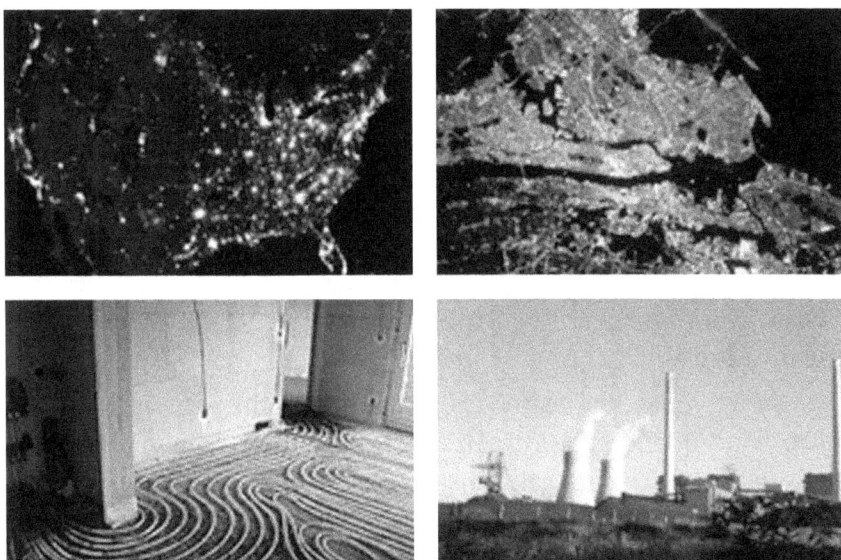
Figure 1.13

The advent of fossil fuels has made life much more comfortable and pleasant for humans. We can be warm in our houses and apartments, and have lots of light, both in our houses and on the streets of our big cities. Seen from space at night (See photo) America displays its brilliant cities, East to West, North to South, across the 48 States. New York City seen from space at Night (See photo) is especially brilliant. Before fossil fuels, viewing America at night would show nothing but a dark plain.

The electric power for today's lights comes primarily from fossil fuels. Power plants – mainly coal and natural gas with a small fraction from nuclear reactors, and a very small fraction from hydro, wind and ground solar. For heating houses, apartments, industrial processes, etc. we rely primarily on oil, natural gas, and electricity to heat water or air, which is then distributed to the heating area by pipes, radiant (See photo) radiators, blowers, etc. No need for fireplaces.

## Conclusion

Of course, we cannot know how readers will answer the three questions posed at the beginning of this chapter. All we can do is give our opinion as to how most of our readers would answer them:

Will humanity give up fossil fuels and today's life style, if it means returning to the life style before we started using fossil fuels, in order to prevent environmental catastrophe?

Answer: No.

How important are future generations? Will humanity give up fossil fuels to protect them from environmental catastrophe, if it means returning to our previous life style? Answer: No.

Sadly, today's concerns are usually more important than long term consequences. *Après Moi Le Deluge*. Also, giving up fossil fuels without a practical substitute for them would result in a massive and terrible extinction for more than 90% of humanity as we returned to the Dark Ages. Which catastrophe is worse?

If new technologies not dependent on fossil fuels can be developed that will maintain and improve our present life styles and not lead to environmental catastrophe, should we implement them? Yes, of course. How soon should they be implemented? As soon as possible.

The above conclusions appear self-evident. If we can avoid environmental catastrophe and then achieve to a much better standard of living by adopting new technologies that do not depend on fossil fuels, humanity should implement them.

In the following 2 chapters, we describe how much fossil fuels the World currently consumes and its effect on the Earth's temperature, weather, food and water supply, and the potential extinction of a large fraction of the fellow species that we share the World with.

# Chapter 2

# Fossil Fuels and Global Warming – The Truths

Many refuse to accept the truths about fossil fuels and what their consumption is doing to Earth's environment. Some can't handle the truth, because they don't want to think about the consequences. Some realize the truth but don't want to acknowledge it for various reasons – it's politically dangerous or to do so would hurt their business and monetary interests or there's money to be made acting as a spokesperson for vested interests.

In this chapter we review the truths about fossil fuel consumption and global warming. In the next chapter we describe the consequences.

Truth #1. Humanity consumes enormous volumes of coal, oil, and natural gas, hundreds of times greater than the volumes of Earth's highest mountains and of humanity itself.

Truth #2. The carbon dioxide ($CO_2$) generated by burning fossil fuels is rapidly increasing the concentration of $CO_2$ in Earth's atmosphere to much higher levels than normal over Earth's 4 Billion year history, and at a much faster rate.

Truth #3. The increasing $CO_2$ level in the atmosphere is rapidly warming the Earth by decreasing Earth's capability to radiate away the energy it absorbs from sunlight. Global temperatures are continuously setting new records.

Truth #4. The rising global temperatures are causing additional $CO_2$ releases from carbonaceous materials present in permafrost regions as they warm up and oxidize. Also, ocean warming is causing marginally stable methane hydrates in sea beds to decompose, releasing methane into the Earth's atmosphere – methane is 20 times more potent than $CO_2$ as a greenhouse gas. At some "trigger point", the warming Earth will experience "runaway global warming", which the stored carbon in the permafrost and sea beds will accelerate and continue to release ever larger amounts of $CO_2$ and methane, even if humans cease consuming fossil fuels.

This environmental catastrophe will wipe out most of the species on Earth, as it did during the Permian Extinction 250 million years ago, when more than 90% of Earth's species went extinct.

Let us now examine Truth #1 in more detail, the immense amounts of fossil fuels we consume. Presenting people with very large numbers of consumption per year, i.e. billions of tons of coal, billions of barrels of oil, zillions of cubic feet of natural gas, is difficult to picture.

A much better approach is to compare the volume of fossil fuels we consume with everyday existing objects on Earth that people see and interact with.

Let's start by visualizing how high the fossil fuel volume consumed per year would be on a well-known area, like Manhattan Island. Figure 2.1 shows a lithograph of Manhattan in 1873, before all those tall building were built on it.

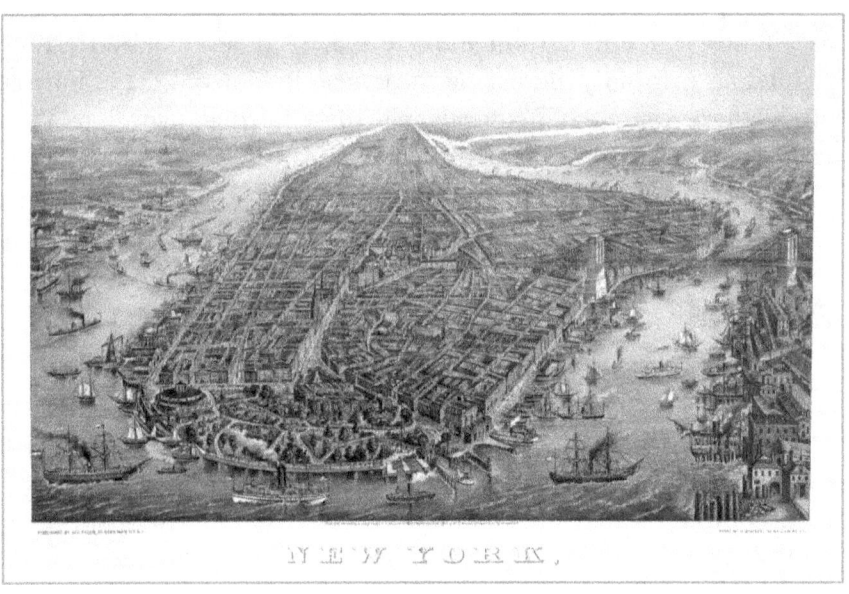

Figure 2.1

Manhattan's land area is 22.6 square miles (59.1 square kilometers). Suppose we dumped 1 year's worth of the World production of coal, oil and natural gas on Manhattan (1). How thick would the pile be?

Figure 2.2 shows the depth of the coal, oil and liquefied natural gas in layers on the entire 22.6 square miles, relative to the 1,378 foot tall World Trade Center #1, before it was destroyed in the 9/11/2001 attack.

The thickness of 1 year's production of coal, oil, and natural gas with a combined height of 1110 feet (337 meters) compared to the 1,368 foot tall (417 meters) World Trade Center. Imagine – a pile of one year's fossil fuel production that is 22.6 square miles in area and 1110 feet high. In terms of volume, that's 20 km³ (cubic kilometers) and that is just one year's production.

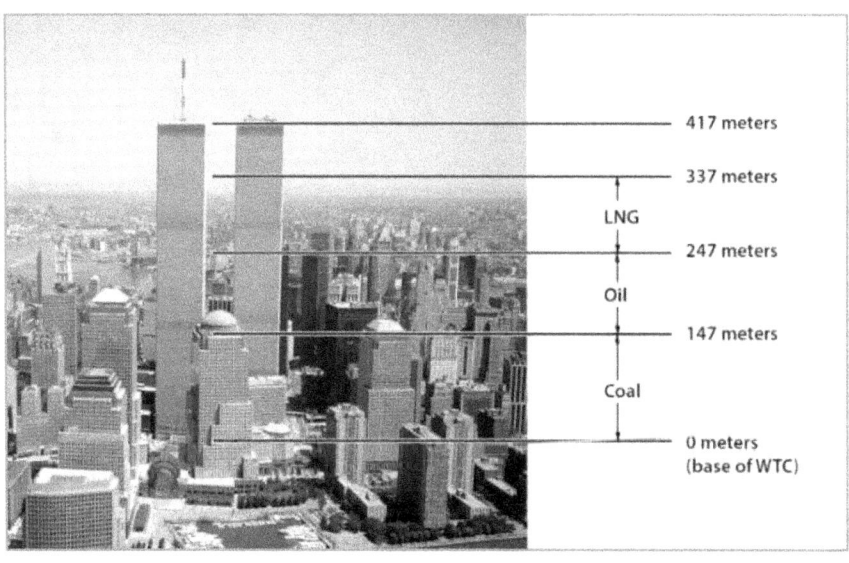

Figure 2.2

Figure 2.3 shows the height of the Manhattan fuel pile as a function of time, starting with 2015. Fossil fuel production rates are assumed to be constant, even though they will grow with time as World population increases and less developed countries industrialize. By 2040, the fuel pile is 5.5 miles high, the height of

Mt. Everest, the highest point on Earth. By 2100, the end of the 21ˢᵗ Century, the fuel pile would be 18 miles high, 3 times the height of Everest.

Another way to visualize the immense volume of fossil fuels we use is to compare their volume with the volume of natural points on Earth. Let's start with the volume of the 7 billion people on Earth today. Taking the average weight of a person to be 100 pounds (45 kilograms), averaged over all the men, women, and children in the World, and a density equal to water, the total volume of the bodies of 7 billion persons is 0.32 cubic kilometers (0.32 Km³).

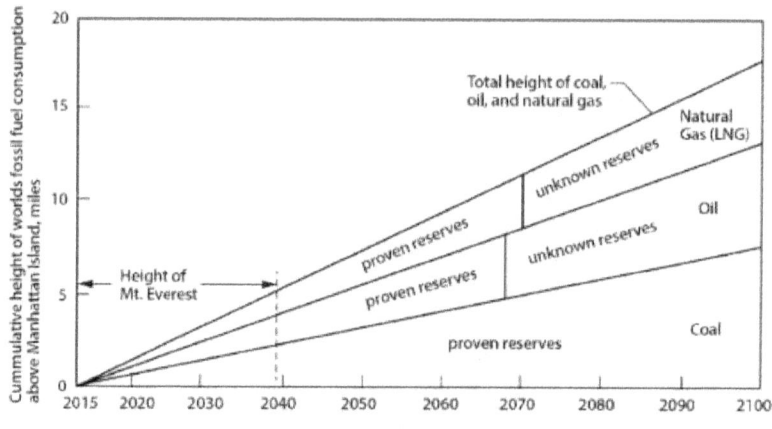

Basis: 22.6 square miles area of Manhattan Island, 2013 world consumption rates for oil, coal, and natural gas

Figure 2.3

The annual World production of fossil fuels is 20 Km³ per year, 60 times greater in volume than the 7 billion people who use the fossil fuels. By 2100 at the current consumption rate of fossil fuels, the total volume consumed will be 85 x 20 = 1700 Km³ ⁻ 5100 times greater than the total volume of 7 billion humans.

Each week, on average, each of us consumes a volume of fossil fuels equal to the volume of our own body.

Another truth about fossil fuels that is not generally realized – we can't keep using them for much longer. We probably will run out of oil and natural gas before 2100 AD, and out of coal not long after that. Figure 2.4 shows the run out dates for oil, natural gas, and coal, assuming the known reserves given in the 2014 BP Statistical Review, for three scenarios:

Present (2014) consumption rates for oil, natural gas, and coal remain constant for as long as reserves last.

World per capita consumption rates for oil, gas, and coal increase to a level of ½ the present consumption rate for Americans. World population stays constant at 7 billion people.

World per capita consumption rate is the same as in scenario 2, i.e. ½ of American per capital consumption rate but World population increases to 9 billion people, the level forecast by the UN for 2050.

| | R, Years Until Known Reserves Run Out | | |
|---|---|---|---|
| Fuel Type | Scenario #1 | Scenario #2 | Scenario #3 |
| | Present World Consumption Rate 7 Billion People (2015) | World Avg Per Capita=1/2 Present US Value 7 Billion People | World Avg Per Capita = 1/2 US Value 9 Billion People |
| Oil | 53 | 22 | 17 |
| Gas | 55 | 22 | 17 |
| Coal | 112 | 78 | 60 |

Figure 2.4

In scenario #1, the World runs out of known reserves of oil and gas by 2070 AD, but still has coal at 2100 AD. In scenarios #2 and #3, we run out of oil and gas much faster, sometime in the period

2035 to 2040 because the less developed countries have industrialized with a much higher per capita consumption rate. We even run out of coal in the 2080 to 2090 period.

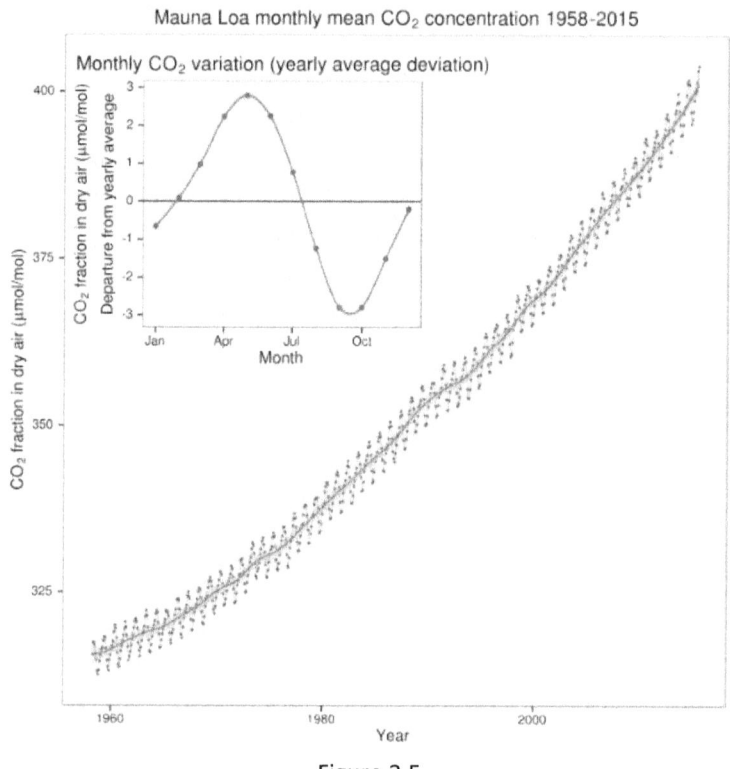

Figure 2.5

Of course new unknown reserves of oil, gas, and coal will be discovered, and the run-out dates will be further in the future. However, there is a lesson here. Even if fossil fuel consumption were not leading humans to environmental catastrophe, fossil fuels are not sustainable in the long term if we keep consuming them to support many billions of people on Earth with an acceptable quality of life. Earth's fossil fuel reserves, known and unknown, will not keep human society going for another 100 years, and probably not even for a shorter time. We must develop massive new energy sources that will not run out and do it quickly.

Now to the 2nd Truth – the impact on Earth's atmosphere of burning immense amounts of fossil fuels, i.e., 20 cubic kilometers of oil, liquefied natural gas, and coal, each year. Figure 2.5 shows the increase in the concentration of carbon dioxide in Earth's atmosphere since 1960, as measured on the Mauna Loa volcano in Hawaii. (2) Called the Keeling curve after the scientist, it shows the carbon dioxide concentration has increased from 300 parts per million (ppm) in 1960 to 400 ppm in 2014.

On a larger time scale starting in 1000 AD the CO2 concentration has increased from 280 ppm to the present 400 ppm, with the increase starting at 1800 AD at the beginning of the Industrial Revolution and the large scale consumption of fossil fuels.

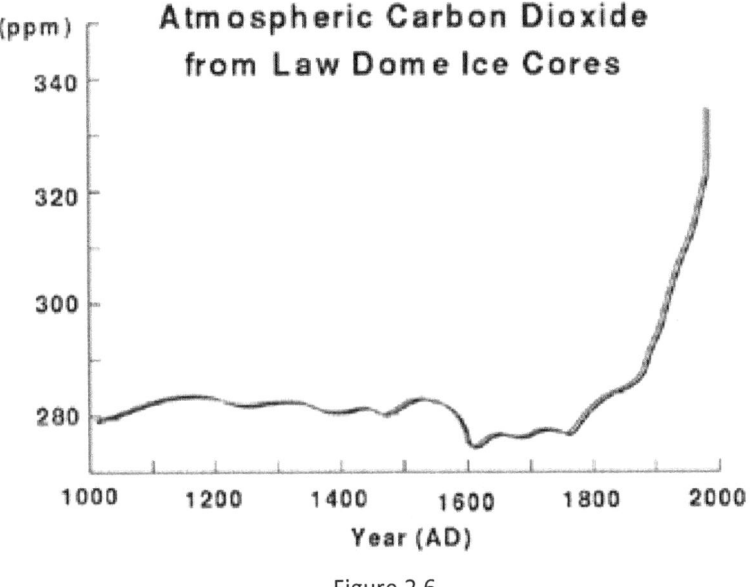

Figure 2.6

The CO2 data in Figure 2.6 were taken from atmospheric air trapped in ice cores extracted at the Law Dome in Antarctica. (3) The Mauna Loa CO2 concentrations match those from Law Dome at the same years. CO2 emitted by burning fossil fuels at one site doesn't stay at the site, but spreads out to all parts of Earth, including Antarctica, producing the same rise in atmospheric concentration everywhere.

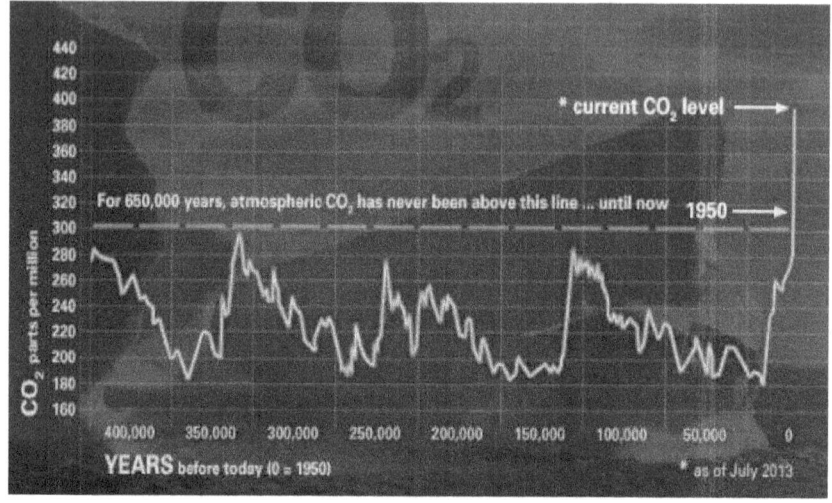

For 650,000 years, atmospheric $CO_2$ has never been above this line ... until now

1950

* current $CO_2$ level

* as of July 2013

Figure 2.7

Today's CO2 concentration of 400 ppm in Earth's atmosphere is far higher than any time in the past 650,000 years as shown in Figure 2.7. (4) The CO2 concentration in past times, measured in Antarctic ice cores, has varied from a low of 180 ppm to a high of 300 ppm. Not only has the CO2 level reached 400 ppm, 100 ppm greater than the previous high of 300 ppm, but it has risen much faster than in the last 650,000 years. The current rate of rise is 100 ppm increase in 60 years, compared to a maximum rate of rise of 100 ppm in 5,000 years in past time. Earth has a much shorter time to adapt to increasing CO2 concentration.

Global warming deniers insist that burning fossil fuels does not warm planet Earth. We discuss, next, Truth #3, that in fact the massive amounts of carbon dioxide released into the atmosphere by burning fossil fuels, i.e. 35 Billion metric tonnes annually, does in fact warm the planet.

Even though the global warming deniers will not accept Truth #3, however, they must accept Truth #2, that humans are emitting

26

tremendous amounts of carbon dioxide into the atmosphere, with the result that the CO2 concentration is rapidly increasing at a rate that is 100 times or more faster than it has increased in the past 650,000 years, and that the CO2 concentration in the atmosphere is 100 ppm higher than it has been in the same period, the result of fossil fuel burning.

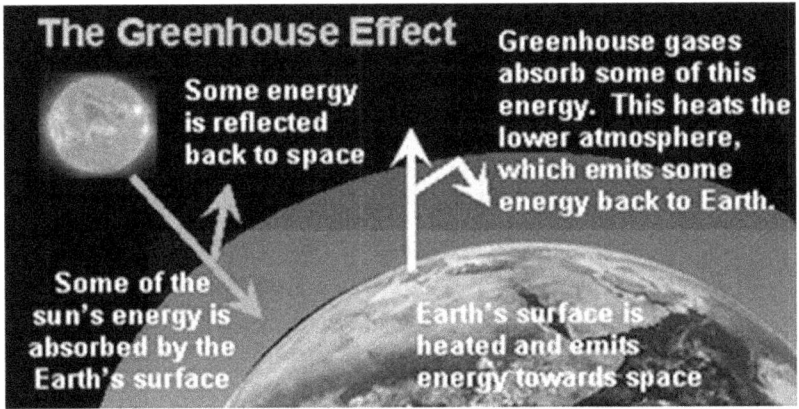

Figure 2.8

To deny these facts would be evidence of incompetent thought or a deliberate lie.

Truth #3, increasing CO2 concentration does make the Earth warmer. The physical process is very simple as illustrated in Figure 2.8. Most of the Sunlight's energy striking the Earth is absorbed by its oceans and land surfaces, with the balance reflected back to space. In turn, Earth has to radiate back to space all of the Sun's energy that it absorbs – otherwise, Earth's temperature would continuously increase and it would melt.

Most of the Sun's energy is in the visible spectrum of light because it is extremely hot, on the order of 6,000 degrees Centigrade, or in absolute temperature, about 6,300 degrees Kelvin. Earth's average temperature is much lower on the order of 27 degrees Centigrade (300 degrees Kelvin, or abbreviated 300 K). As a result, virtually all of the energy that Earth radiates to space is in the infrared spectrum of radiant energy, the same wavelength of energy that one feels standing next to an electric heater.

As Earth's radiant energy travels through the atmosphere it gets absorbed by certain gases in the atmosphere. The energy absorbed in these gases is then re-emitted. 50% of the re-emitted energy continues as radiation into space, while 50% of the energy is radiated back to Earth's surface, where it is absorbed making the Earth warmer. These gases, termed "greenhouse gases" because of their warming effect include carbon dioxide, methane, etc., with carbon dioxide a principal component.

The more carbon dioxide in the atmosphere, the greater its absorbing power, and the warmer the Earth becomes. The process is familiar to chemical engineering sophomores in college, when they analyze radiant heat transfer in combustion furnaces. The greater the concentration of carbon dioxide in the combustion gas layer around a thermally radiating surface, and the thickness of the gas layer, the higher the temperature of the radiating surface. The sophomores have the advantage of using the "Bible" of chemical engineers, Perry's Chemical Engineering Handbook. Inside is a nice chart that gives the emissivity of gases containing carbon dioxide as a function of its partial pressure, i.e., concentration, the length L of the gas, for a combustion gas pressure of 1 atmosphere. Data from the chart is shown in Figure 2.9, for the condition that the radiating surface and the combustion gas layer are sufficiently close in temperature that the gas emissivity and absorptivity are essentially the same.

The sophomore then can calculate the temperature of the radiating surface knowing the energy input and the concentration of carbon dioxide in the gas layer. As an old chemical engineer (James Powell), I can say, "Bless Perry's Chemical Engineering Handbook."

Figure 2.9
Absorptivity of Thermal Radiation in air as a Function of Carbon Dioxide Partial
Pressure and Radiation Path Length

| $P_cL$ (atmft) | Data Source | Carbon Dioxide Concentration (ppm) | Absorptivity of Thermal Radiation |
|---|---|---|---|
| 0.2 | Perry | ---- | 0.10 |
| 1 | Perry | ---- | 0.15 |
| 5 | Perry | ---- | 0.20 |
| 7.2 | Extrapolation of Perry Data | 280 ppm (pre-industrial | 0.215 |
| 9.77 | " | 380 ppm (present) | 0.225 |
| 12.34 | " | 480 ppm (future) | 0.235 |
| 14.34 | " | 580 ppm (future) | 0.245 |

Basis: Data from Perry's Chemical Engineers Handbook, 6th Edition, McGraw-Hill (1984)

$P_c$ = Partial Pressure of $CO_2$ (atm) in 1 atm air

L = Radiation Path Length

(25,700 feet in Earth's Atmosphere) normalized to constant pressure of 1 atm)

Air Temperature = 530 Degrees Rankine (70F)

Figure 2.10 shows geologic data on global temperature to 2000. The temperature rise tracks closely the atmospheric CO2 concentration measured at Mauna Loa (Figure 2.5) from 1960 to 2015 and at Vostok (Figure 2.6), from 1000 to 2000 AD.

The same correlation between atmospheric CO2 concentrations and global temperature over the last 650,000 years is shown in Figure 2.10. When CO2 concentrations go up, so does global temperature. When CO2 concentrations go down, so does global temperature.

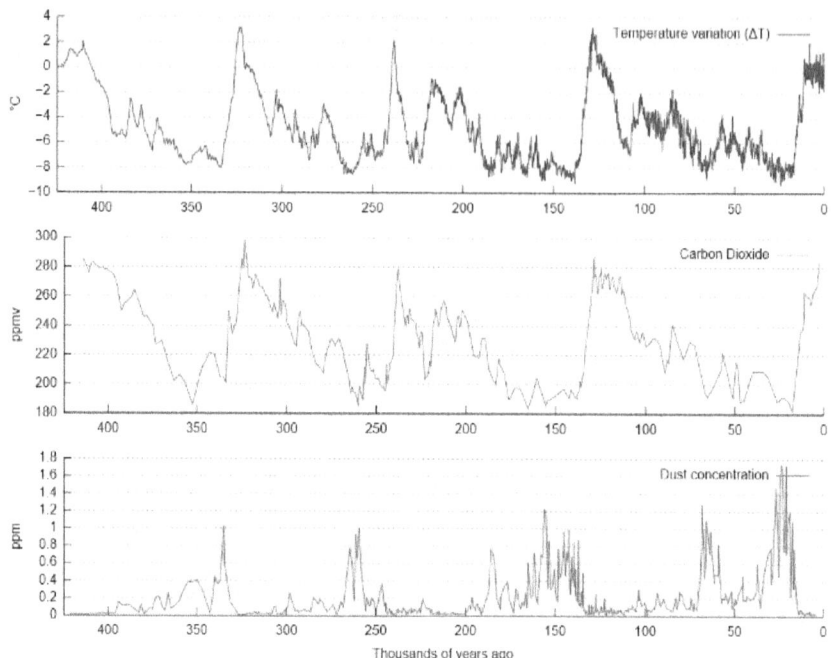

Figure 2.10

## Energy change inventory, 1971-2010

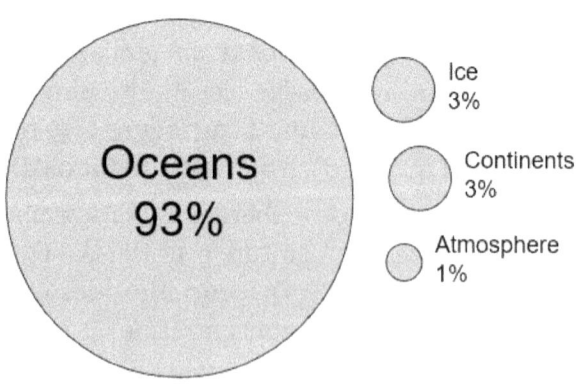

Figure 2.11   Energy Change Inventory, 1971-2010

Figure 2.11 shows the distribution of the energy trapped by greenhouse gases. (2) Most, over 90 percent, ends up in the oceans, because it is a much more effective heat sink than the continents, atmosphere, and ice. First, oceans occupy ¾ of Earth's surface area. Second, they have a much greater heat capacity and a greater rate of heat transfer through them than land or ice. The surface of a sand desert has a much lower rate of transferring heat into the ground beneath it, compared to a liquid water surface, where convection currents can transfer energy at a much faster rate to the water beneath the surface.

Increasing ocean temperatures and $CO_2$ absorption will have profound effects on ocean life and food chains, as discussed in the next chapter.

We now come to the final truth, #4. It would be comforting to believe that humans can control the level of $CO_2$ in the atmosphere and the temperature rise that goes with increasing concentrations. At some point in the future, when humans finally realize the environmental damage that fossil fuels have caused and decide to stop, then the environmental damage will stop getting worse, and we will just have to live with it. Right? Not necessarily.

Figure 2.12

Heat transfer does not stop if we stop burning fossil fuels. Thermal energy in warm surfaces will conduct down through the soil beneath, increasing the temperature of the cooler soil, even after the burning ceases.

There is an enormous amount of carbonaceous material frozen inside the permafrost soil in the Arctic regions, in the range of 1700 to 1850 billion tons of carbon, more than double the amount presently in Earth's atmosphere. (5) As it warms and oxidizes it will release carbon dioxide and methane into the atmosphere, methane is 20 times more potent than carbon dioxide as a greenhouse gas. Figures 2.12 and 2.13 show pictures of the permafrost surface and interior.

Figure 2.13

As the permafrost warms, at some point, termed the "Trigger Point" its increasing release of carbon dioxide and methane will become great enough that atmospheric greenhouse gas concentration and global temperature will increase exponentially, even if humans completely stop consuming fossil fuels.

Even greater sources of stored carbon that could be triggered for an exponential increase of greenhouse gases are the deposits of marginally stable methane deposited on the sea floors. At high pressure and cold temperatures, gaseous methane condenses inside ice, forming a solid structure. The structure is marginally stable. If warmed, it will release its methane, which can burn, as shown in Figure 2.14.

Estimates of the ocean's methane hydrate inventory are in the range of 500 to 2,500 billion tons of carbon compared to a total of 800 billion metric tons of carbon in the atmosphere. (6)

Released as methane into the atmosphere into the atmosphere, where methane is more than 20 times more potent than carbon dioxide, 500 billion tons of carbon in methane would be equivalent in greenhouse gas warming capability to 20 x 500/800 or 12 times greater than the 400 ppm of carbon dioxide already in the atmosphere. Earth would fry, like eggs on a hot griddle.

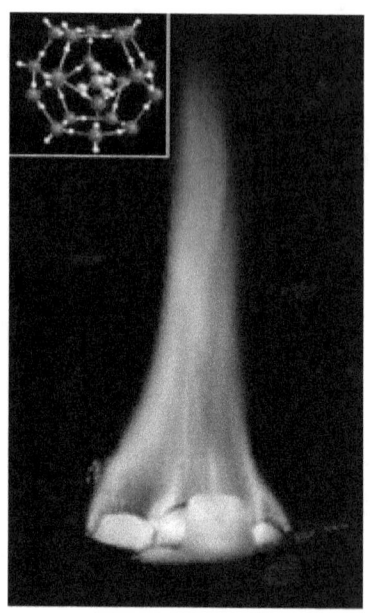

Figure 2.14

Methane boils in the ocean and permafrost have already been observed. At his point their contribution to global warming is small, relative to the much greater effects from fossil fuel consumption. However, as the Earth continues to warm, methane and CO2 emissions from oceanic methane clathrates and the frozen organic material in the permafrost will inevitably increase.

When will the methane and CO2 releases reach the Trigger Point, and global temperature increase exponentially, even if humans stop burning fossil fuels? We don't know. Nobody knows. We may have already passed it. But we can't wait until we know for sure, because we will have already passed it and it will be too late. For a sustainable long-term human civilization, we must start reducing fossil fuel consumption as soon as possible, and at the fastest possible rate of reduction.

# Chapter 3

# Fossil Fuels and Global Warming – The Consequence

*"For that which is common to the greatest number has the least care bestowed upon it. Everyone thinks chiefly of his own, hardly at all of the common interest; and only when he is himself concerned as an individual. For besides other considerations, everybody is more inclined to neglect the duty which he expects another to fulfill; as in families many attendants are often less useful than a few."*

Aristotle, Politics, Book II, Chapter III

In 1968, Garret Hardin published his classic essay, *"The Tragedy of the Commons"*. (1) Hardin focused on how unfettered access to resources by individuals not responsible for sustaining the resources which he termed the "commons", inevitably leads to their overuse and degradation. While the observation has been made before, by Aristotle and others, Hardin expresses it very succinctly and forcefully. To prevent this from happening, and the unsustainable growth in World population that the "Commons" has enabled, resources should be controlled and regulated, and human population growth constrained.

Examples of "The Tragedy of the Commons" are all around us. Rainforests are cut down (Figure 3.1) for growing food, animal grazing, palm oil production, etc. Millions of acres of rainforest have been destroyed around the World, causing species extinction, loss of soil fertility, habitat destruction, etc. Farmers apply fertilizers and pesticides, polluting water supplies of millions of people living far away. Companies dump PCB's and other dangerous pollutants into rivers, with no concern about their effects and people living down river. Fluorocarbon emissions destroy the World's protective ozone layer, and on and on.

35

The effects of fossil fuel consumption are an incredible "Tragedy of the Commons" and a devastating attack of Earth's environment. The previous chapter focused on the effects of the carbon dioxide emissions from burning fossil fuels on the $CO_2$ concentration in the atmosphere and the resultant increase in global temperature.

Figure 3.1 Cutting and Burning Rain Forest

The consequences of increasing global temperature are already very serious, and will be much more serious in the years ahead. The consequences include:

- Massive melting of sea ice, polar ice caps, and glaciers
- Rising sea levels and coast flooding
- Severe storms and storm surges
- Droughts, wildfires, and crop failures
- High temperature heat wave deaths
- Increase in diseases and blights

36

Ocean acidification and species extinction addition to the environmental devastation from increasing global temperatures, the other wastes released by producing and burning fossil fuels include:

- coal mine tailings, pollution of rivers and aquifers
- large dumps of toxic coal ash from burning coal that pollutes water supplies
- mercury, nitrous oxide, and other pollutants released into the atmosphere and water supplies
- micro particulates from fossil fuel consumption released into the atmosphere that damage hearts and lungs
- pollution of aquifers from drilling for oil and gas
- oil spills in ocean waters from drilling rigs and oil tankers

In this chapter, we focus on the consequences of increasing global temperatures. They are humanity's primary concern, particularly with respect to runaway global warming and ocean acidification, which would lead to the 6th extinction in Earth's history, in which most species, including probably humans, would go extinct. The time scale is short for humanity. If we continue consuming fossil fuels at our present rate, in a few decades we will pass the "Trigger Point", with runaway global warming and the death of much of ocean life.

The "Tragedy of the Commons" lets us escape from immediately realizing the consequences of our fossil fuel consumption, because we dump the carbon dioxide and other greenhouse gases into Earth's vast atmosphere. We don't experience the consequences in the short term. The very long term consequences? That's a problem for future generations. "After me, the Deluge". Polluted aquifers? Massive dumps of toxic coal ash and other hazardous wastes? That's someone else's problem .Our attitudes on fossil fuel consumption would radically change if there were no commons to dump our carbon dioxide and other waste products into. Let us visualize an imaginary tribe of humans called the "Manhattanites". The Manhattanites live on a small island with high population density (Figure 3.2). The island is 22.8 square miles (59.1 square kilometers) in area, with 1.6 million inhabitants.

Figure 3.2 Midtown Manhattan

They consume, per capita, the same amount of oil, coal and natural gas as the average American does today, and discharge, per capita, the same amount of $CO_2$ and other greenhouse gases, along with coal ash and other wastes as the average American. The big difference is that the Manhattanites do not have a global commons to discharge their carbon dioxide and other wastes into. To paraphrase the old saying, "whatever happens in Las Vegas, stays in Las Vegas", is discharged by Manhattanites, stays in Manhattan.

What are the consequences to Manhattanites? Let's consider 3 waste products:

- $CO_2$ into the air above Manhattan Island.
- Coal ash from the coal they burn
- Water released by the melting of glaciers and polar ice caps per amount of fossil fuel consumed

All of the above stay in Manhattan. First, what is the rate of melt water rise on the island? Today, the Greenland ice and Antarctic ice sheets are melting at combined rate of 500 cubic kilometers of ice per year. (BBC News) Adding in glacier melt, that is over 500 $Km^3$ of melt water per year. With a world population of 7 Billion people, the per capita melt rate is 70 cubic meters (18,000 gallons) annually.

For the Manhattanites, that corresponds to a water level rise of 1.9 meters (6 feet) per year in the island. In 10 years that would be 60 feet – can't ignore it and say that it's a problem for future generations.

Per capita, Americans burn approximately 3 metric tons per year of coal, resulting in about 0.3 metric tons of coal ash. The 1.6 million Manhattanites would generate 500,000 tons of coal ash annually, with a volume of approximately 1 million cubic meters per year. That would be a ½ inch thick layer added to Manhattan's surface area, if spread out uniformly across the island. Though messy and harmful to health, the Manhattanites could survive the ash for a good period.

The real killer for the Manhattanites is the $CO_2$ emissions. America emits 5.4 Billion tons of carbon dioxide per year from burning fossil fuels, 18 tonnes per American. Total $CO_2$ emissions for the 1.6 Manhattanites is 30 million tonnes annually.

Released into the atmosphere above Manhattan island, that corresponds to a concentration increase in air of 33,000 parts per million (ppm) per year, 1,000 ppm every 11 days. The lethal concentration of $CO_2$ in air is 100,000 ppm (2) so all the Manhattanites would be dead in 3 years. However, they would be dead much sooner from the terrible heat waves caused by high $CO_2$ concentrations in the atmosphere. At 1000 ppm, two and one-half times greater concentration than Earth's present $CO_2$ concentration of 400 ppm, every Manhattanite would die of heat stroke. Two weeks to live, unless they discharge the products of fossil fuel consumption to the Global Commons. Would they hesitate to dump everything into the Global Commons? Are you kidding?

In 1962, Rachel Carson (Figure 3.3) published her groundbreaking book, "Silent Spring" (3)(4) Her book brought to the public's attention the destructive effects of pesticides on humans and wildlife particularly birds. Her book and public appearances led to a ban on the use of DDT for agriculture applications and the formation of the Environmental Protection Agency.

Figure 3.3 Rachel Carson

The indiscriminate use of pesticides is a good example of the "Tragedy of the Commons". Farmers apply DDT and other pesticides to increase crop yields without concern about the resulting decline in bird populations in the commons.

Despite very strong opposition from the chemical industry and agricultural interests, including threats of law suits, Rachel Carson and Silent Spring succeeded in curtailing the use of pesticides and making the public aware of their destructive effects on the environment.

## Why was Carson successful?

For 3 reasons:

- The consequences of indiscriminate pesticide use on bird populations could be seen in real time, not in the far future.
- The public cared about birds. They were their everyday experience and liked them very much.
- The consequences to the public of banning DDT and other pesticides were very minor. Their standard of living did not suffer.

For fossil fuel consumption, however, the situation is very different. First, the really massive consequences of burning fossil fuels are decades in the future, so that people can put off making decisions, or deny that there will be consequences. Bird population decline from pesticides is today. The 6th extinction and runaway global warming? Let future generations deal with it, if it really happens.

Second, the consequences of not consuming fossil fuels are very bad – it would return us to the Dark Ages, as described in Chapter 1. That is why it will be impossible to give up fossil fuels, unless new technologies can take their place that maintain, or even better, improve humanity's quality of life. In Part 2 of this book, we describe how Maglev, a fundamentally new mode of transport can cleanly move people and goods much faster, better, cheaper, and safer than existing transport systems, how it can launch space based solar power systems into orbit that will beam down very low-cost, environmentally clean power to Earth, and how synthetic gasoline and diesel fuel and other carbon based materials, e.g., plastics, etc., can be cleanly manufactured from air and water.

We believe that this approach, providing humanity with new technologies that do not require fossil fuels and that will substantially improve our quality of life, is the best, and really the only approach, that can lead to humanity giving us fossil fuels.

Now, we came back to a description of the consequences of global warming that are occurring as a result of the growing concentration of carbon dioxide and other greenhouse gases emitted by the consumption of fossil fuels. We start with:

## Massive Melting of Sea Ice, Polar Ice Caps, and Glaciers

Average temperatures in the Arctic region around the North Pole have increased substantially in the last 3 decades. Figure 3.4 shows much of the Arctic region increasing by approximately 3 degrees Celsius (5.4 degrees Fahrenheit) in the year October 2010-September 2011, compared to the long-term average from 1981 to 2010.

The rising temperatures are melting sea ice and the Greenland ice sheet. Figure 3.5 shows melt ponds in the Arctic sea ice sheet, while Figure 3.6 shows how much the minimum area covered by Arctic sea ice has shrunk in 2012 from its larger extent in 1984. By 2050, it is projected that the Arctic sea ice will disappear completely during the warmer months. Good luck, polar bears!

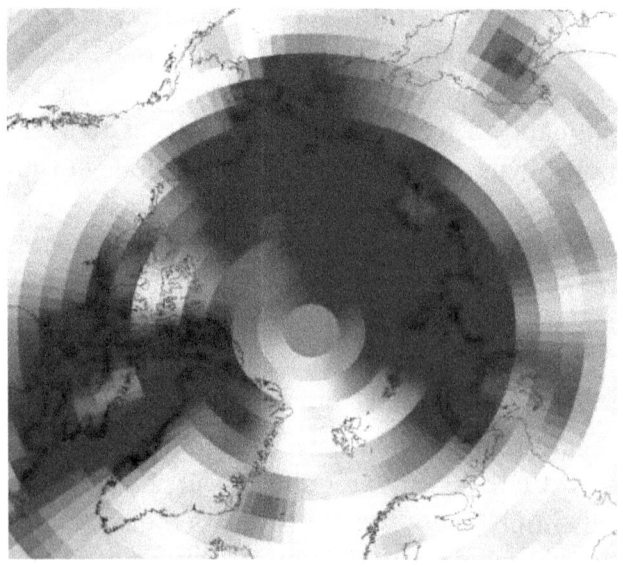

Figure 3.4 Temperature Rise in the Arctic Region

Figure 3.5 Melt Ponds in Arctic Sea Ice

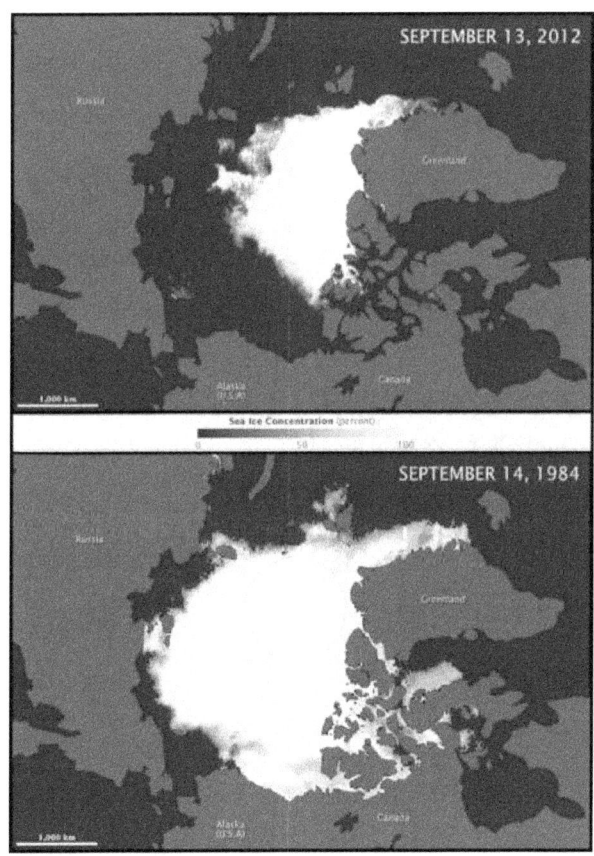

Figure 3.6 Sea Ice Area In the Arctic Ocean, 1984 and 2012

The Greenland Ice Sheet shown in Figure 3.7 and the Antarctic Ice Sheet in Figure 3.8 are also melting, discharging large volumes of ice into the oceans. Figure 3.9A shows an iceberg calved from the Greenland Ice sheet, while Figure 3.9B shows a portion of the Ross Ice Shelf, about to slide into the ocean.

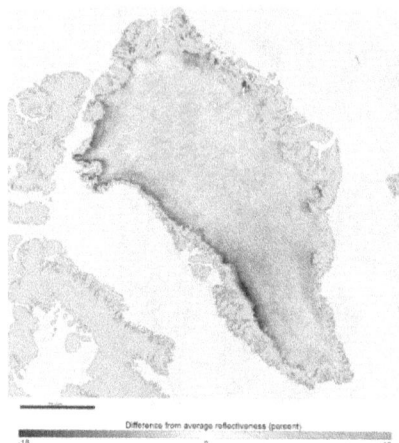

Figure 3.7 Greenland Ice Sheet

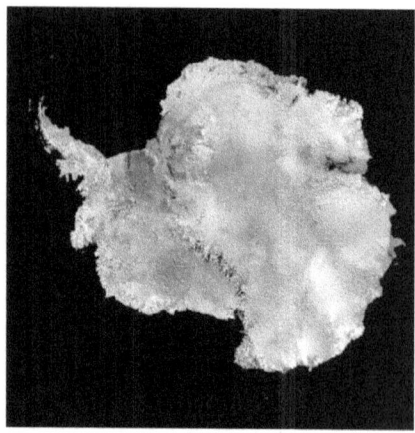

Figure 3.8 Antarctic Ice Sheet

Figure 3.9A Iceberg Calved from Greenland

Figure 3.9B Portion of Antarctica's Ross Ice Shelf About to Slide into Ocean

There is an immense amount of ice on Greenland, approx.-mately 2.8 million cubic kilometers (680,000 cubic miles). If it were all to melt, sea level would rise by 7.2 meters (24 feet) (5).

Before global warming, the ice sheet was in balance, with no net loss of ice. Every year, the volume of new snow that fell on Greenland was balanced by the volume of icebergs calved from it. No problem, except for ships like the Titanic.

With global warming, however, there is now a net loss of 375 cubic kilometers of ice per year (6) and it is accelerating. Several factors are contributing to make the ice sheet move more rapidly towards the coast of Greenland, losing a greater volume of ice and making it thinner. Melting water at the base of the ice sheet reduces friction and the thinner ice sheet is more mobile. Also, melt water on the surface of the ice sheet, shown in the map of Greenland (Figure 3.7) decreases the albedo of the ice surface. With a decreased albedo, less of the sunlight striking the ice sheet is reflected back into space, and more of it is absorbed, further warming the ice sheet. Some areas of the ice sheet are reflecting 20% less sunlight than they did 20 years ago. (5)

The enormous volume of ice on Greenland is dwarfed by the even greater volume in Antarctica (Figure 3.8), 26.5 million cubic kilometers (6.36 million cubic miles), approximately 10 times the ice volume on Greenland. (7)

If it were all to melt, sea level would rise by 70 meters (230 feet). Most everybody's home would then really be underwater.

The annual net loss of ice from Antarctica is 128 cubic kilometers, about 1/3 of the 375 cubic kilometers from Greenland (5) even though its ice mass is much greater. But it will accelerate. Of particular concern is the West Antarctic ice sheet, which if it collapses would raise sea level by 3.3 meters (11 feet) in a short time. (8)

Finally, glaciers are rapidly melting in all of the continents and countries around the World, in the tropics, the temperate zone, and high latitude regions. Figure 3.10A shows the extent in 1973 of the White Chuck glacier, in Washington State, while Figure 3.10B shows the shrunken glacier in 2006. The Mer de Glace in Mt. Blanc, the largest glacier in France, has retreated 500 meters (1600 feet) between 1994 and 2008. (9)

Figure 3.10A Whitechuck Glacier State of Washington, 1973

Figure 3.10B Whitechuck Glacier 2006

In Alaska, 99 percent of the glaciers are retreating. The Columbia glacier near Valdez has retreated 15 kilometers (9.3

miles) in the last 25 years. In Africa, the glacier on top of Mt Kilimanjaro has retreated 75% since 1912, losing 80% of its mass. At the present rate it will be gone by 2020 AD. Almost all of the glaciers in the US Glacier National Park will be gone by 2030 AD. By 2050 virtually all of the World's glaciers will have vanished. What a shame –all that natural beauty gone.

The ice melted from the glaciers and the ice sheets on Antarctica and Greenland flow into the ocean, raising sea levels and causing coastal flooding around the World, the next consequence of global warming.

## Rising Sea Levels and Coastal Flooding

Global sea levels are rising. Figure 3.11 shows measurements of the global sea level rise, about 9 inches, from 1880 to the present.

The rate of sea level rise is accelerating with time. Depending on the source, there is a range of predictions as to how much sea level will rise by 2100, the end of the 21st Century. The 2013 IPCC (International Panel on Climate Change) forecasts in its worst case scenario that sea level would rise by an additional meter (3 feet) by 2100 AD. (8)  In the lowest greenhouse gas emissions scenario sea level would rise by 0.3 to 0.6 meter (1 to 2 feet) by 2100. The Third National Climate Assessment (NCA) projected a sea level rise of 0.3 to 1.2 meters (1 to 4 feet) by 2100. Other assessments project even greater sea level rises by 2100, as much as 2 meters (6.6 feet)

What coastal areas would be affected by rising sea levels? Figure 3.12, a 2012 NASA map shows the coast areas and cities affected by a 6-meter sea level rise. Many of the World's island nations, e.g. the Maldives and others, would completely vanish into the sea.

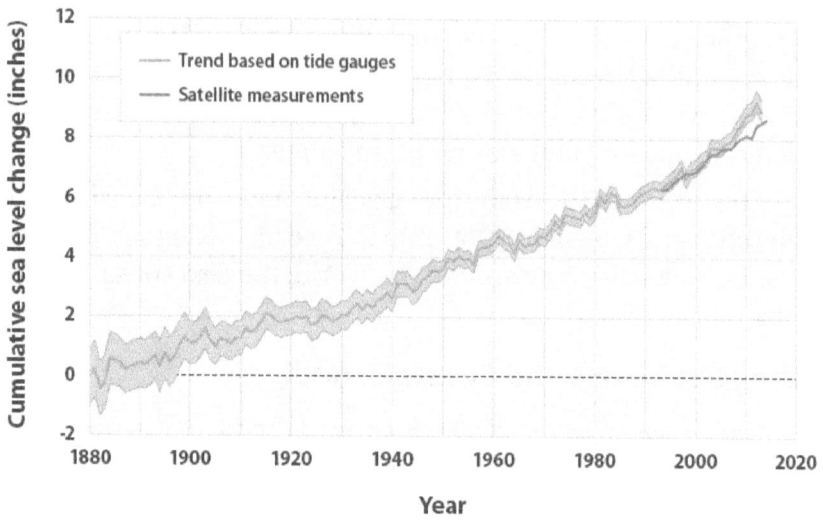

Global Average Absolute Sea Level Change, 1880–2014

Data sources:
• CSIRO (Commonwealth Scientific and Industrial Research Organisation). 2015 update to data originally published in: Church, J.A., and N.J. White. 2011. Sea-level rise from the late 19th to the early 21st century. Surv. Geophys. 32:585–602. www.cmar.csiro.au/sealevel/sl_data_cmar.html.
• NOAA (National Oceanic and Atmospheric Administration). 2015. Laboratory for Satellite Altimetry: Sea level rise. Accessed June 2015. http://ibis.grdl.noaa.gov/SAT/SeaLevelRise/LSA_SLR_timeseries_global.php.

Figure 3.11 Global Average Sea Level, 1880-2014

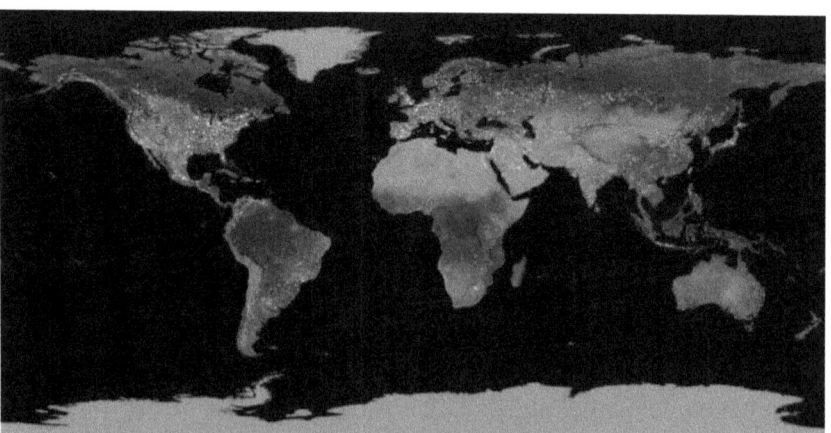

Figure 3.12 NASA World Map of Effect of 6 Meter Sea Level Rise

Studies estimate (8) that approximately 600 million people live in coastal areas that are less than 9 meters (30 feet) above sea level, and that 2/3 of the World's cities that have 5 million or more inhabitants are located in those coastal areas.

The insurance company SwissRe estimates that the Southeast area of Florida will have an economic loss of 30 Billion dollars from climate - related damages. Miami is ranked as "the number-one most vulnerable city world-wide" to damage from storm surges and sea level rise.

Figure 3.13 shows a map by EPA of the areas that would be flooded for 3 scenarios of sea level rise – 1, 3, and 6 meters, and the resulting effect on existing power plans that would be impacted by the rising sea levels. A substantial fraction of the power plants would have to shut down. (10)

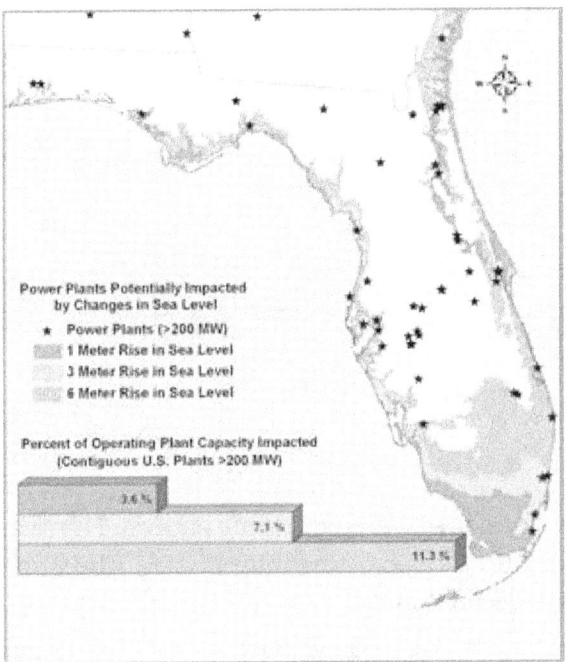

Figure 3.13

Much of the population in Asia and Africa live in low coastal areas. The 2007 IPCC Report estimated that a sea level rise of only 0.4 meters in the Bay of Bengal would put 11 percent of

Bangladesh's coastal land underwater, displacing 7 to 10 million people (8).

Global warming not only causes sea levels to rise, it also makes storms and storm surges stronger, increasing the flooding and damage to populations in coastal areas from hurricanes and severe storms.

## Severe Storms and Storm Surges

Rising global temperatures make Earth's oceans and land surfaces hotter. Water evaporates faster, and its concentration in the atmosphere increases. There's more thermal energy for forming storms, and they increase in intensity. Stronger winds, more rain, higher surges of ocean water pushed ashore during hurricanes, greater flooding, more loss of lives, and increased damage to coastal populations.

Table 3.0

Analysis of hurricane intensity over the period 1970 to 2005 (10) yielded the following results with a doubling in the fraction of hurricanes in Categories 4 and 5, the most severe.

Time Period and Fraction of Hurricanes in Category

| Hurricane Intensity | 1920-1974 | 2000-2004 |
|---|---|---|
| Category 1 | 43% | 30% |
| Category 2 and 3 | 40% | 33% |
| Category 4 and 5 | 17% | 37% |
| | 100% | 100% |

Other studies have found the same pattern. Hurricane intensity is increasing due to global warming and will become even more intense in the decades ahead.

Hurricanes also termed cyclones – and their associated storm surges can cause tremendous damage and loss of life in coastal regions. The 1970 Bhola Category 3 cyclone which struck Bangladesh and West Bengal killed 500,000 people. In the most affected area, Upazila, Tazumuddin, more that 45% of the 167,000 population died. (11)

In 2008 Cyclone Nargis killed more than 138,000 people in Myanmar.

America has had its share of deadly hurricanes. The 1900 Galveston Category 4 hurricane wiped out most of Galveston, Texas, killing more than 6,000 People. More recently, we have experienced Hurricane Katrina in 2005 and Hurricane Sandy in 2012. (12)

Figure 3.14 shows a satellite photo of Hurricane Katrina as it approached New Orleans and the Gulf Cast. Katrina was a Category 5 hurricane in the Gulf of Mexico, but weakened to Category 3 at its landfall on the Gulf Coast. Even though it had weakened, the damage was enormous, not only in New Orleans, but in neighboring cities and area on the Gulf Coast.

Figure 3.14 Hurricane Katrina Aug. 28, 2005

80% of New Orleans was flooded (Figure 3.15). Storm Surge height was 8 meters (25 feet). Boats were swept inland, (Figure

51

3.16) to be left on land after the storm surge receded. 1,836 people died.

| Figure 3.15 Flooded New Orleans | Figure 3.16 Boats Swept Inland by Storm Surge |

Figure 3.17 Hurricane Ike Gilchrist damage 22 September 2008

After Hurricane Katrina came Hurricane Ike in 2008, a Category 4 hurricane, which also hit the Gulf Coast, causing tremendous damage. Figure 3.17 shows the damage in Gilchrist,

Texas, and Figure 3.18 the devastation to the Bolivar Peninsula in Texas.

Figure 3.18 Damage caused by Hurricane Ike in the Bolivar Peninsula, Texas Sept 2008

Then came Hurricane Sandy. It hit Cuba as a Category 3 hurricane, (12) Moving North, it weakened but continued to do enormous damage. On October 29th, 2012, it hit New Jersey as a post tropical cyclone with hurricane force winds.

Figure 3.19 shows a satellite view of Hurricane Sandy approaching the US East Coast. 233 people died from Sandy. 157 in the US, with 71 of those in New York State. Total damage was $65 Billion. (12)

Figure 3.19 Hurricane Sandy NASA satellite image

Figure 3.20 shows the Sandy damage to the Long Beach area on Long Island. The East River in New York City overflowed, flooding much of Manhattan. The storm surge was 14 feet high at Battery Park. Subway tunnels under the East River were flooded along with station power lines. Sandy damaged or destroyed 100,000 homes on Long Island, with more than 2,000 homes non restorable.

As global temperatures increase in the years ahead, hurricanes will become much stronger, and the fraction in categories 4 and 5 will continue to increase. Storm surge heights will increase, along with greater numbers of deaths and much worse economic damage.

Figure 3.20 Sandy Damage to Long Beach Island.

Storms and storm surges do their damage with too much water. Next, we describe the damage done by global warming when it reduces the amounts of water available to certain regions on Earth – droughts, wildfires, and crop failures.

## Droughts, Wildfires, and Crop Failures

Global warming is causing severe droughts in many parts of the World. As the droughts continue, forests and vegetation shrivel and die, making wildfires much bigger and more numerous. People in drought areas have less water for agriculture, industry, and personal needs. Soils dry up and crops fail, causing starvation and famine in some areas.

The western US is already experiencing very extreme droughts over much of its area, with predictions that the droughts will become much worse in the decades ahead. Snow packs in the mountains – the source of much of its water – are disappearing. Reservoirs and lakes such as the Salton Sea and others, are also disappearing. Inhabitants in the drought areas are meeting their water needs by pumping out their aquifers, drilling ever deeper for water. In many regions, to reach water, one has to drill down hundreds of feet. As the aquifers are exhausted, the ground level goes down in some places as much as 30 feet, compacting soil so that it cannot replenish its water if rain does come. At some point, the aquifers will be completely exhausted, and desertification will ensue.

Figure 3.21 shows a US Drought Monitor Maps of California on October 21, 2014. 58% of the State is in the D-4 exceptional drought condition, and 92% in the D-3 and D-4, extreme drought and exceptional drought condition.

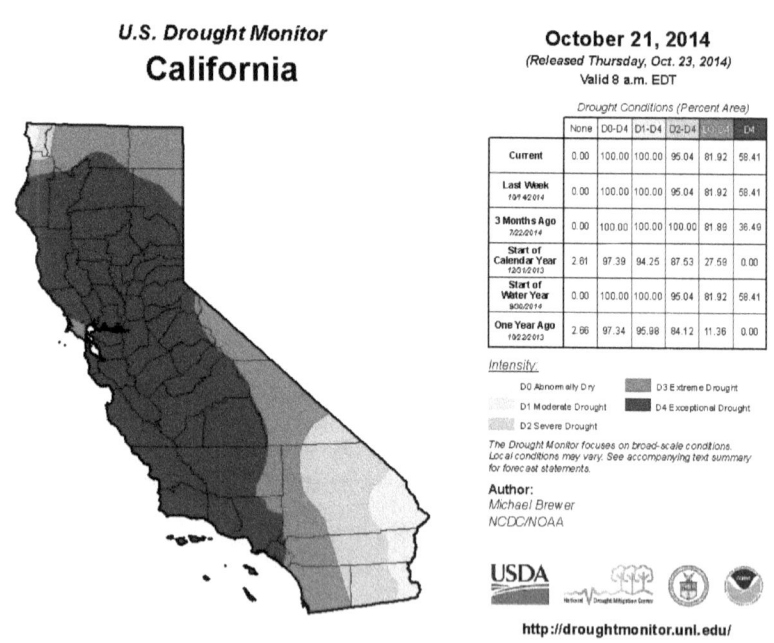

Figure 3.21 California Drought Status Oct 21, 2014

Figure 3.22 shows the US Drought monitor Map for the Western US on August 25, 2015, a year later. California continues in the D-3 and D-4 categories, extreme and exceptional drought, with D-3 & D-4 droughts in parts of Nevada, Utah, Oregon, Washington, Idaho, and Montana. Other states have areas of severe drought. Figure 3.23 shows President Obama and Governor Brown of California visiting a drought area and talking to a California farmer and his wife.

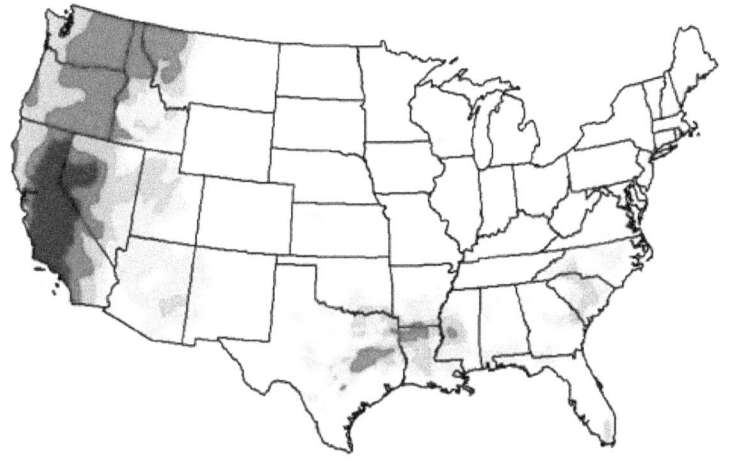

Figure 3.22 US Drought Monitor Sept 15, 2015

Severe droughts are occurring on all of the World's populated continents – North and South America, Asia, Africa, Europe, and Australia. They will intensify as global temperature increase.

Wildfires will also intensify. In the Western US and Alaska very large land areas are burning. Figure 3.24 shows a satellite view of wildfires in Southern California and Baha California on May 27, 2014, while Figure 3.25 shows a satellite view of the smoke from wild fires in Northern California on August 17, 2015.

The Western wildfires are very intense, with the flames higher than the trees that are burning, for the 2008 California wildfire shown in Figure 3.26. Following the fire, only charred dead trees and vegetation remain, as shown in Figure 3.27 for a forest fire in Washington State.

California wildfires burned 580,000 acres of land in 2013 and 630,000 acres in 2014. Each year, about ½ percent of California's land burns in wildfires. And the fires are getting more intense and burning more area as global warming intensifies. The costs, both human and economic, to California are tremendous.

Figure 3.23 Governor Brown and President Obama on a visit with California Farmers

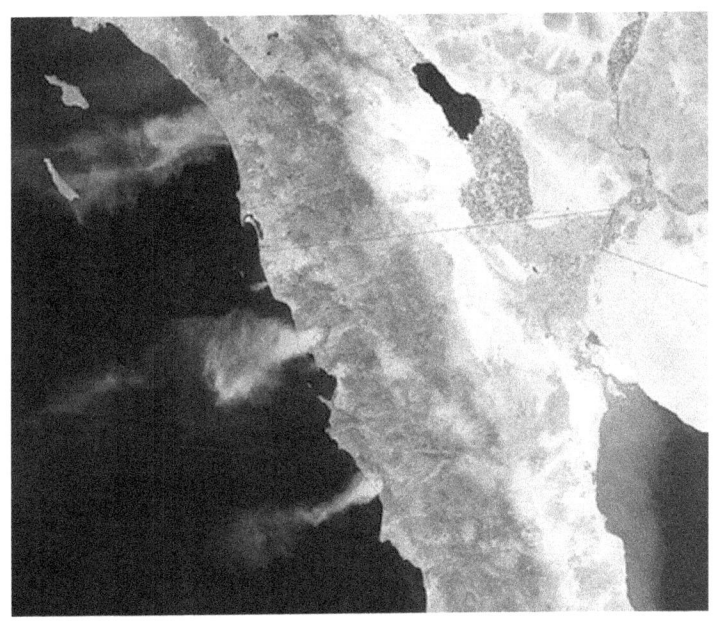

Figure 3.24 Satellite View of California Wildfires

Figure 3.25 Northern California

Figure 3.26                    Figure 3.27 Aftermath of Forest Fire in
                                                    Washington State


Wildfires are not limited to California, but occur over a large portion of the Western US and Canada. Figure 3.28, shows a NASA map of wildfire intensify in the Western US and Canada over the period 2000-2013.

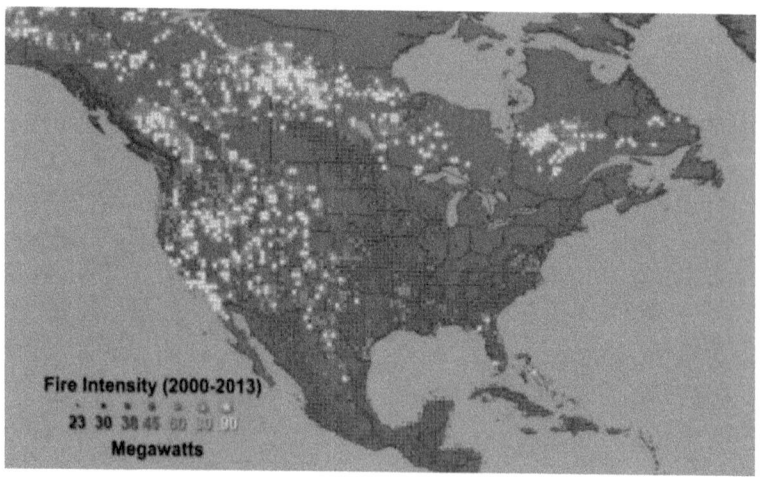

Figure 3.28 Wildfire Intensity 2000-2013

How will global warming affect the World's agricultural output? This is a complex question. Droughts will certainly cause large reductions in crop yields and animal production. One can't grow fruits, grains, vegetables, and animal feeds without water. However, other factors are also important. If temperature levels rise too far, plants will grow more slowly and beyond a certain temperature not produce seed. Corn will not reproduce at temperatures above 95 degrees Fahrenheit. Soybeans will not reproduce at temperatures above 102 degrees Fahrenheit. (13)

Most of the World will see substantial reductions in crop yields as global temperatures increase. There may be some gains in yield for certain crops grown in high latitudes as their temperatures increase, but overall, there will be major reductions. Figure 3.29 shows the decrease in global crop yields as a function of global temperature rise for African maize, Asian rice, India wheat, US maize, and US corn. (13)

Assuming that the global temperature rise can be held to 2 degrees Celsius, yields for the 5 crops would decrease on average about 20%. With a 4 degrees Celsius rise, a very possible scenario, the average decrease would be about 50%. There would be massive starvation, famine, and population decline.

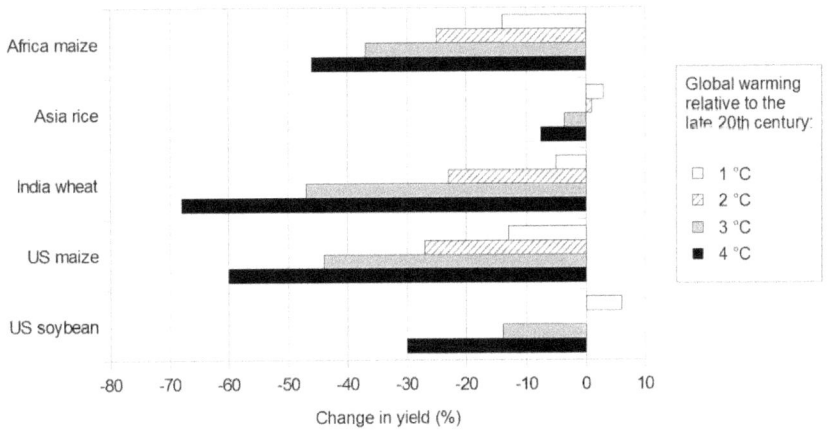

Figure 3.29 Projected Changes in yields of selected crops with global warming

61

The above projections do not include possible additional reductions due to increasing drought and the impact of longer population of pests that attack crops.

High temperatures cause crops to stop growing and reproduce. They also cause humans to sicken and die, the next consequence of global warming we consider.

## Humans and Heat Waves

In 2003, Europe had the hottest summer on record since 1540, with France the hottest country. Total heat caused deaths in Europe were more than 70,000, with 14,800 occurring in France. (14)

Figure 3.30 shows a map of the temperature increase across Europe during the July 20 to August 20, 2003 period from the average over the same period for 2000, 2001, 2002, and 2004.

There were record temperatures in France (104°F.), Portugal (118°F.), Spain (113°F.), Italy (115°F.), Germany (106°F.), Switzerland (107°F.), and United Kingdom (101°F.). (14)

Deaths caused by heat stress (hyperthermia) are not limited to Europe, but occur world-wide – in the US. Africa, India, China, etc., affecting most of the World's population. The 1995 Chicago heat wave, for example, caused 600 deaths over a period of 5 days. (15)

Heat stress and hyperthermia depend not only in temperature, but also the relative humidity in the air we breathe. The heat stress is measured by the heat index – the higher the humidity at a given temperature, the higher the heat index and the greater the damage to humans.

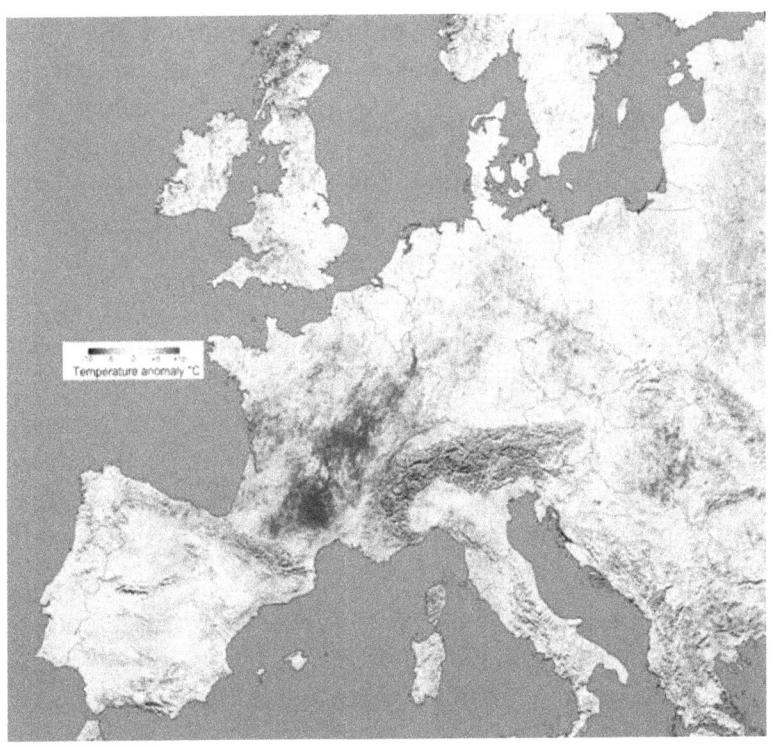
Figure 3.30
Canicule Europe 2003

The NOAA National Weather Service measures the index as a function of the 2 variables, temperature and relative humidity. At a heat index of approximately 130, persons are in extreme danger of hyperthermia. (15).

The heat index at 40% relative humidity and 110°F is 136, extreme danger. A 100% relative humidity and 92°F, it is 132, again extreme danger. In between, at 100°F and 65% relative humidity, the heat index is 136, extreme danger. So one can endure higher temperatures in dry air than one can in very humid air, but still, once temperature rises above 110°F, a person is probably in trouble. As global temperatures rise, temperatures of 118°F or greater are becoming much more common in the World, including the US.

While high temperatures are bad for humans they tend to be good for diseases, pests, and blights, the next consequence.

## Diseases, Pests, and Blight

Human diseases love warmer, wetter climates. Mosquitos carry malaria, elephantiasis, Rift Valley Fever, yellow fever, dengue fever and the rapidly spreading Zirka virus. (16)

Ticks also spread disease, including Lyme disease, and their populations are rapidly growing as global temperatures rise. Other diseases on the rise include Hantavirus, schistosomiasis, river blindness and tuberculosis. Cholera is also likely to increase as flooding increases and the cholera bacteria infiltrates drinking water. Toxic algae blooms increase as ocean temperatures increase. While not infecting humans, the do infect fish and shellfish, killing humans when they eat the seafood.

Then there are the pests that attack vegetation and food crops. As of May 2013, the Pine Beetle (Figure 3.31) has killed 88 million acres of forest in 19 western US States and Canada, at a 70-90% kill rate. (17)

Global warming lets pine beetles breed more rapidly, increasing their population. There are many other kinds of pests that attack forests and crops, whose population will increase as global temperatures increase.

Crop blights will also increase as global temperatures increase, due to the faster and greater growth of bacteria and other diseases that attack crops. Figure 3.32 shows an example of fire blight that has attacked a Gala apple tree.

Blights are caused by bacteria, molds and fringe. They can infect potato, rice, corn, etc. plants, as well as apple, pear, chestnut, etc. trees.

So far, we have focused on the consequences of global warming on life forms –humans, plants, and animals that live on Earth's land surfaces. We now turn to the consequences for life in the oceans, ocean acidification, and species extinction.

5 millimeters

Figure 3.31 Pine Beetle

Figure 3.32 Severe Fire Blight
Infection in Gala apples

## Ocean Acidification and Species Extinction

A large fraction, about 30 to 40% of the 35 Billon tonnes of fossil fuels emitted by burning fossil fuels, doesn't stay in the atmosphere, but is absorbed in Earth's oceans. This is good news for the atmosphere as it helps to slow down the rise in temperatures. It's bad news for the oceans, because it makes them more acid. In the period from 1751 to 1994, the H+ ion concentration in seawater, a measure of ocean acidity, increased 30%. (18)

Increasing acidity adversely affects many marine organisms. It reduces the concentration of carbonate ions in the seawater, available for forming shells on many organisms including mollusks, crustaceans, reef forming corals, and some species of algae and phytoplankton. (19)

The concentration of carbonate ions in the oceans is expected to decline by a factor of 2 by the end of the century, making it harder for many marine organisms to form shells. Moreover, since the oceans will be more acidic their shells will dissolve as fast as they try to form them. The effect of increasing ocean activity on shell formation can be studied and measured in the laboratory (18). In much of the oceans, it is projected that shell formation will not be possible.

Inability to form shells will not only affect the marine organisms that depend on shells, but also the organisms that that feed on them in food chains. If organism A depends on eating organism B, and B goes extinct, so will organism A.

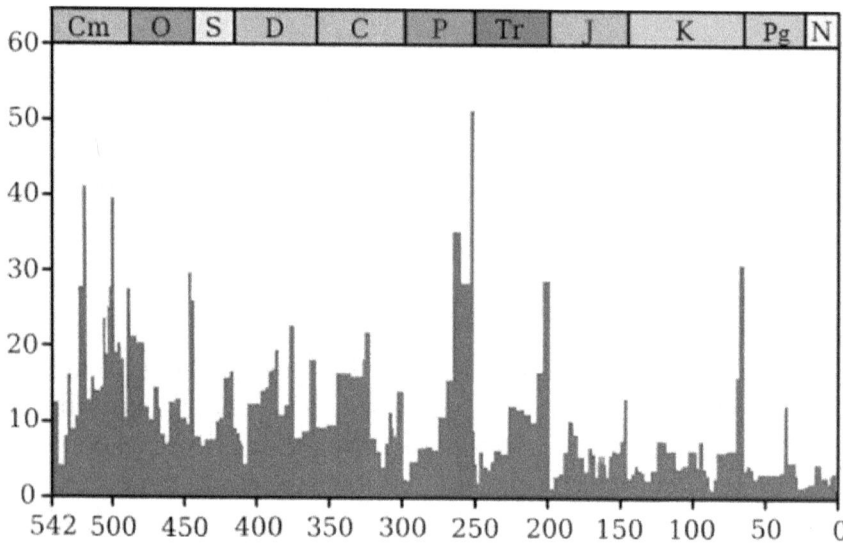

Time runs from left to right (millions of years ago). Vertical axis is percent of species lost.

Figure 3.33 Extinction Intensity

In the Arctic, for example, acidification threatens to destroy major food chains. Pteropods and brittle stars are at the base of the food chain for larger organisms – larger plankton, fish, seabirds, and whales. (18) Wiping out the pteropods and brittle stars could wipe out the organisms higher up on the food chain. Instead of vibrant oceanic life, we could have oceanic deserts by 2100 AD, if fossil fuel consumption continues at current rates.

Humanity is already at the beginning of the "Sixth Extinction", in which a large fraction of Earth's species will go extinct in a short time. Figure 3.33 shows a graph of the previous 5 mass extinction 250 million years ago, when 90% of Earth's species went extinct. (20)

Humanity has already driven many species into extinction through habitat destruction, e.g. cutting down rain forests, overhunting, and over fishing. Figures 3.34 and 3.35 show paintings of the Dodo and Passenger Pigeon, extinct because of overhunting. Lions, tigers, and elephants are now on the road to extinction. Large animals going extinct get a lot of attention but there are many smaller animals that have already gone extinct without attracting any attention. As an example, Figure 3.36 shows a photo of the Golden Toad, which was last seen in 1989.

The IPCC (International Panel on Climate Change) 4th Assessment Report projects that 20 to 30 percent of the species assessed will be at risk of extinction if global temperatures increase by more than 1.5 to 2.5 degrees Celsius relative to 1980-1999 temperatures. As average global temperatures exceed 3.5 degrees Celsius, projections suggest 40 to 70 percent of the species assessed will go extinct. (21) The current rate of extinction is 10 to 100 times faster than the natural rate.

We have only looked at a portion of the consequences of global warming that is being caused by massive consumption of fossil fuels. Including the additional consequences not covered, the environmental outlook is even more disastrous and dangerous than depicted here. And when the Trigger Point is passed and run-away global warming occurs, Earth will experience environmental catastrophe.

Figure 3.34 Edwards' Dodo bird

Figure 3.35 Passenger Pigeon

Figure 3.36 Golden Toad

## Our Fossil Fuel Future: Will it be the Sirens and the Rocks or The Raft of the Medusa?

If humanity continues to consume vast quantities of fossil fuels, there will inevitably be environmental catastrophe. World population will shrink to the level in the Dark Ages or lower, with a radical decrease in the standard of living, like that described in Chapter 1, or even worse.

Will the descent into a new Dark Age take place quickly, or will it be drawn out agony over decades? Will it be like the voyagers in ancient Greek mythology, who when hearing the beautiful songs of the Sirens, shown in the paintings, were so attracted to them that they crashed on the rocks and died. Suffering and death, but it was short.

Odysseus had his sailors plug their ears with beeswax so they would not hear the Sirens song and crash his ship. But he was tied to the mast, and could hear and enjoy their wonderful songs with no danger to him and his crew. Wise man.

In our time, the Sirens are not beautiful women with wondrous songs, but fossil fuels. Not as attractive perhaps, but they do promise automobiles, televisions, fast foods, movies, warm and comfortable houses, good lighting, cellphones, computers, and the Internet, etc.

But the rocks lie ahead. The oceans become acid and turn into deserts, major pandemics break out, much worse than the 1918 flu epidemic, massive famines due to sudden crop failures, and new World Wars occur, including nuclear weapons use, as countries fight over the rapidly dwindling World Resources.

Figure 3.37 Sirens

Or will the catastrophe be like the Raft of the Medusa, depicted in Gericault's glorious painting?

The raft of the Medusa was a real and tragic event. The French naval frigate Medusa ran aground off the coast of Mauritania on July 2, 1816. (22) 147 people were set adrift on a raft, but only 15 were still alive when they were rescued 13 days after the accident. During the 13 days, the other 132 persons on the raft died of starvation, dehydration, being killed and eaten or thrown overboard by the survivors, and suicide. When their ordeal began, their supplies for 147 persons were only one bag of ships biscuits, 2 casks of water (lost overboard during fighting) and a few casks of wine.

So instead of human civilization quickly crashing on the rocks when fossil fuel environmental catastrophe suddenly happens, it could decline over decades with population levels declining as resources dwindled. There would be conflicts and killings, starvation as crops failed, floods, storms and the other consequences described in Chapter 3, but no World Wars and no major pandemics, just increased rates of disease.

Figure 3.38 Gericault's Raft of Medusa

Which of the two paths will we follow, if humans continue to use fossil fuels at the present rate? Nobody knows, or can know at this point. All we know is Yogi Berra's quote, "The future lies ahead".

# Part II

# How Maglev Can Help Stop Global Warming

*Much have I travell'd in the realms of gold,*
*And many goodly states and kingdoms seen;*
*Round many western islands have I been*
*Which bards in fealty to Apollo hold.*
*Oft of one wide expanse had I been told*
*That deep-brow'd Homer ruled as his demesne;*
*Yet did I never breathe its pure serene*
*Till I heard Chapman speak out loud and bold:*
*Then felt I like some watcher of the skies*
*When a new planet swims into his ken;*
*Or like stout Cortez when with eagle eyes*
*He star'd at the Pacific—and all his men*
*Look'd at each other with a wild surmise—*
*Silent, upon a peak in Darien.*

*On First Looking into Chapman's Homer*
*BY JOHN KEATS 1795–1821*

72

# Chapter 4

# Maglev Transport for America and the World

Over the ages, humanity has evolved ever faster, better, and cheaper modes of transport. We have gone from hunter-gatherers who walked in search of food and water, to horses, camels, and ships, to wheeled wagons drawn by horses and oxen, to wheeled trains and wheeled motor vehicles.

We escaped from the wheel with the invention of balloons by the Montgolfier brothers and airplanes by the Wright brothers.

Modern transport is wonderful. We can cross continents in a few hours and fly to other continents. Americans travel 15,000 miles per year on average, over 80 percent by auto, to work, to shop, to see friends, for entertainment and vacations. Ships, trucks and trains bring us the food and goods we need to stay alive.

Modern transport is indispensable. Without it we would be living like our ancestors did hundreds and thousands of years ago, as described in Chapter 1.

But we cannot go on consuming vast amounts of fossil fuels for our modern modes of transport. We need a new mode of transport – Maglev. Maglev, short for MAGnetic LEVitation, has no wheels and no rolling friction with rails and roads, Maglev vehicles are magnetically levitated and propelled along guideways at high speeds, hundreds of miles per hour, limited only by air drag. They are electrically powered – no fossil fuels -- with greater energy efficiency than existing modes of transport.

Maglev transport was a far-out dream of Robert Goddard, the American pioneer of rocket transport, and other scientists and engineers, until James Powell and Gordon Danby published their 1966 breakthrough paper on Superconducting Maglev. Their paper sparked Maglev development programs in many countries, in particular, Japan and Germany.

Japan has developed Powell and Danby's 1966 Maglev inventions in their 1st Generation Maglev System, now operating in Yamanashi, Japan (Figure 4.1). The demonstration route has carried over 100,000 passengers, at speeds up to 360 mph with an accumulated running distance of hundreds of thousands of kilometers.

Japan plans to extend the Yamanashi line to become a 300-mile route between Tokyo and Osaka, which will carry 100,000 passengers daily with a trip time of 1 hour.

Figure 4.1 Photo of 1st Generation Japanese Superconducting Maglev Operating at Yamanashi Test Facility

Powell and Danby have continued working to evolve 1st Generation Maglev to an even better, more capable, 2nd Generation System much like airplanes went from the 1st propeller driven Ford Tri-Motor and DC-3 in the 1930's to today's modern jet airliners – Boeing's 767, Airbus'380, etc.

Detailed descriptions of the 2nd Generation Maglev 2000 System, the envisioned National Maglev Network, and Maglev Public Transit are given in two books, "The Fight for Maglev" and "Maglev America", by James Powell, Gordon Danby and James Jordan. The books are available at Amazon.com.

The 2nd Generation Maglev System has the following unique capabilities not possessed by 1st Generation Maglev, that are critically important for intercity Maglev Networks and Maglev Public Transit in congested urban areas.

The 2nd Generation Maglev can:

- Carry trucks and autos as well as passengers (Figures 4.2 and 4.3). Intercity truck revenues are much greater than intercity rail passenger revenues, more than $300 Billion annually, in the US, compared to 2 Billion annually for Amtrak rail passengers. The much greater revenues will enable private financing, with no government subsidies.

- Operate in levitated mode on **existing** rail and subway tracks that have been adapted at very low cost for Maglev – a capability not possible for 1st Generation Maglev Systems. This capability enables 2nd Generation Maglev vehicles to transition from 300 mph guideways to operate at lower speed on **existing** rail tracks when they come into suburban area. This eliminates the need to construct new guideways in suburban and urban areas – an expensive and disruptive process, like Boston's Big Dig. Conventional trains and subway cars can still use the existing tracks after the adaptation process, sharing them with the Maglev vehicles, given appropriate scheduling.

The conversion process is simple and low in cost. Panels of ordinary aluminum loops are attached to the crossties of the existing railroad or subway tracks. As the Maglev vehicle passes over them the magnetic interaction between the superconducting magnets on the vehicle and the aluminum loops on the guideway, maintains its levitation. AC current in a second set of aluminum loops on the crossties magnetically propels the vehicle, providing whatever speed is desired – acceleration, constant speed, and deceleration.

The cost of conversion to Maglev is small, about 10 million dollars per 2-way mile. Maglev operation on existing commuter rail lines, like the 700-mile Long Island Rail Road (Figure 4.4) would enable much shorter trip times, more convenient service, lower fares, and reduced government subsidies.

Electronically switch at high speed from the main guideway line to offline stations for loading/unloading. This eliminates mechanical switches, the method used in 1st generation Maglev. Mechanical switching at low speed significantly reduces the average speed of Maglev vehicles on the main line.

To load/unload passengers, autos, or trucks, 2nd Generation Maglev vehicles will electronically switch to a secondary guideway that leads to the station. After loading/unloading, the 2nd Gen Maglev vehicle than accelerates, switches back to the mainline at high speed and heads for the next stop. No cumbersome, low speed mechanical switches. 2nd Generation Maglev vehicles can by-pass stations at full speed that they are not scheduled to stop at. 2nd Gen Maglev vehicles can travel at full speed as individual units directly to their destinations, without having to stop at multiple stations along the route, a necessity for long trains pulled by locomotives.

Erect elevated high speed guideways at much lower cost than 1st Generation Maglev. The 2nd Gen Maglev monorail guideway beams and piers can be mass produced in factories with all their attached equipment, and then transported at very low cost by highway trucks to the construction site, to be rapidly erected by conventional cranes on pre-poured footings. Very little field construction work is required for 2nd generation Maglev, which is typically much more expensive than factory production. 1st Gen Maglev systems require much more field construction work. The minimal field construction requirements, plus the small footprint of 2nd generation Guideways, enables 2nd Gen Maglev systems to be easily erected on existing rights-of-ways along highways. (Figure 4.5) power transmission corridors and other routes without disruption of the activities that operate there.

Figure 4.2
Artist's depiction of loading a Maglev roll-on, roll-off Truck or Van carrier

Figure 4.3
Artist's depiction of a Maglev auto carrier loading a double-deck carrier.

With cost efficient, mass production of Maglev components – beams, piers, aluminum loop panels, propulsion loops, and electronic as integrated units in factories that can be trucked to sites, the construction cost per 2-way mile of the 2$^{nd}$ Generation Maglev-2000 elevated guideways is only $30 million per 2-way mile. This does not include the cost of Maglev vehicles, stations, and any required land modifications.

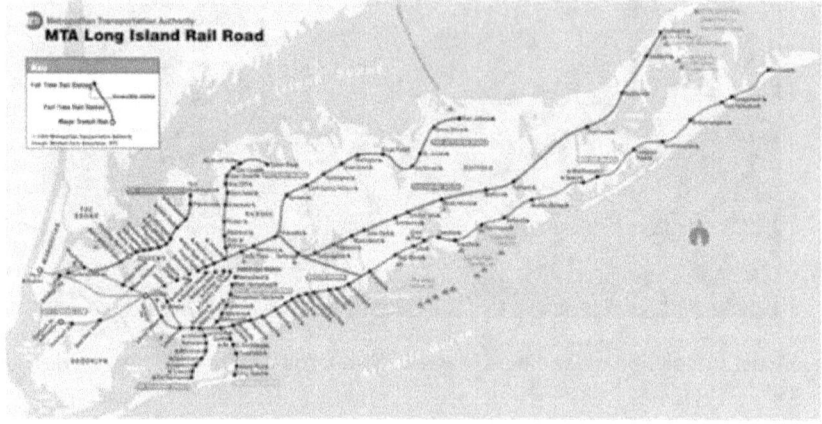

Figure 4.4

The cost of adapting existing rail and subway tracks for Maglev operations is very low, about 10 million dollars per 2-way mile. The panels containing aluminum loops for levitation and propulsion are attached to the crossties of the existing trackage during periods of low travel by the existing conventional trains or subway cars. Field construction is not required.

The minimization of field construction, besides reducing costs, also enables both Maglev high speed elevated guideways for intercity transport and moderate speed Maglev public transport to be installed much more rapidly that systems that require substantial field construction. This is very important in rapidly implementing intercity Maglev Networks and Maglev Public Transit.

The National Maglev Network would be built in 3 waves, each wave taking 5 years. The First Wave would be constructed on the US East and West Coasts where most Americans live.

The World's present fossil fuel powered transport systems are not sustainable and will not be able to meet our future needs.

Within this Century, known oil reserves will be exhausted and the increasing carbon dioxide concentration in the atmosphere will result in global environmental disaster, unless humanity dramatically reduces its dependence on fossil fuels.

Figure 4.5
Depiction of Interstate Highway Rights-of-Way being used for Maglev guideways operating with both a freight carrier to left and passenger vehicle to the right side of the artist's concept .

The realities of the World's present transport systems are very sobering – the increasing amounts of transport required to sustain humanity's economy and standard of living, the incredible number of horrible deaths and injuries on the World's highways, the trillions of dollars spent annually on transport, the growing congestion and travel delays, the effect of the greenhouse gases on global warming, the millions of persons whose health is damaged by pollutants emitted by autos, trucks, ships and other transport systems using fossil fuels, the tremendous environmental damage from the mining and processing of fossil fuels, and so on.

The solutions presently proposed for long term sustainable non-fossil fuel transport– biofuels, hydrogen cars, electric autos, and high speed rail are not practical and will not meet the World's future transport needs. Biofuels compete with food production and could only satisfy a tiny fraction of our transport fuel needs. Hydrogen cars would require enormous amounts of electric power to manufacture the hydrogen, and would be extremely dangerous to drive on the highways. Electric cars are limited in range, expensive, and pose reliability problems for large-scale operations on highways. High speed rail (HSR) systems operate now, but travel costs are very high, and require major government subsidies for construction and operation. Even in countries with HSR systems, they only supply a small fraction, less than 10%, of the countries' transport needs. In America, because of its much large geographic size and much lower population density than in countries that have HSR systems like Japan, China, France and the rest of Europe, HSR would play an even smaller role, providing much less than 1% of US passenger mile transport.

Biofuels are not practical because they conflict with food supply. To think that biofuels will solve our dependence on fossil oil and help to mitigate global warming is to not recognize reality. Follow the numbers. 40% of the US corn crop was sold in 2011 to make 13 Billion gallons of ethanol fuel. (1) It takes 1.5 gallons of ethanol to have the same fuel energy as 1 gallon of gasoline, so that 13.9 Billion gallons of ethanol actually equals only 9 Billion gallons of gasoline. The US consumes 200 Billion gallons of gasoline, diesel, and jet fuel annually, so 40% of its corn crop yields less than 5% of our transport fuel needs.

How much farmland would it take to produce the equivalent of 200 Billion gallons of oil based fuel? A lot. At 400 gallons of ethanol per acre (2), equivalent to 260 gallons of gasoline, it would take 200 Billion/260, or 770 million acres. Total US farmland is 300 million acres, only 1/3rd of the required 770 million acres. So if we want to use ethanol for vehicle fuel, we can stop eating. However, we'll only be able to travel 1/3rd of the distance we travel today.

Clearly, biofuels cannot meet the World's transport needs. In a World of more than a Billion starving and malnourished people who need food, it is heartless to use food to make fuel for automobiles.

Along with Biofuels enthusiasm a few years ago, there was a lot of excitement about hydrogen cars. "Only water comes out of the tailpipe!" "Hydrogen fuel cells are very efficient!" and so on. Unfortunately there are 2 fundamental problems with hydrogen cars – where do we get the hydrogen, and safety. First, free hydrogen doesn't exist in nature. It must either be manufactured from fossil fuels, i.e., natural gas, as in the production of hydrogen for ammonia fertilizer, or produced by electrolyzing water using massive amounts of electric power.

To use fossil fuels to make hydrogen for autos would be incredibly stupid. To make it by electrolyzing water using electric power from fossil fueled power plants, which today generate more than 80 percent of America's power is also incredibly stupid. It would greatly increase fossil fuel consumption and greenhouse gas emissions.

The electric power for hydrogen fuel for motor vehicles would have to come from nuclear and renewable sources – wind, solar, hydro, etc. A hydrogen transport economy would need a lot of power. Based on EPA analyses, electric cars use approximately 0.36 Kwh of electric power per mile of travel from their battery packs with power transmission and distribution losses and battery inefficiencies, extracting 0.36 Kwh per mile from an electric car battery pack would require generating about 0.5 Kwh/mile at the electric power plant.

The electric power required per mile for fueling hydrogen cars would considerably greater, due to power losses in electrolyzing water to make the hydrogen, compressing the hydrogen to 5,000 psi for on-board storage in cars or liquefying it to very low temperatures, 20 degrees Kelvin, for storage of liquid hydrogen in your car tank, plus the losses in the hydrogen fuel cells on the car that generate electric power for operating it. All told, the primary electric power generated for a hydrogen car will be approximately 1.0 Kwh/mile, about 2 times that for an electric car.

Total passenger vehicle miles in the US (autos, SUV's, light trucks in 2011 were 2.6 Trillion. Total US electric generation was

3.9 Trillion Kwh. If all 230 million cars were hydrogen powered, American would have to boost its power generation by 2.6 Trillion Kwh, a 66% increase – equivalent to adding 370,000 megawatts of wind, ground solar, or nuclear power – that's 370 new 1,000 megawatt nuclear reactor power plants.

Plus the safety issue. Visualize 230 million hydrogen cars driving 70 mph on America's congested highways, with each car holding either a tank of compressed hydrogen gas at 5,000 psi, or a cryogenic tank of liquid hydrogen at 20 degrees Kelvin. In a collision, if the hydrogen escapes into the atmosphere, it could explode. For a hydrogen fueled car with a 300 mile range, the explosive force would be equivalent to many pounds of TNT.

For the reasons given above, electric cars will probably be the choice for much of America's motor vehicle transport needs in a non-fossil-fuel future. However, they do have significant limitations.

- limited range
- run out of power on highway
- long charging times
- availability of charging stations
- high cost

The Chevy Volt has an EPA range of 35 miles and the Nissan Leaf, 75 miles. The Tesla Model S EPA range is 265 miles with an 85 Kwh battery pack. However, these ranges relate to ideal conditions – not in cold weather where heating of the vehicle interior would be necessary, not in hot weather where air conditioning is necessary, or on congested highways where there are long delays. Non-ideal conditions will substantially reduce electric car range. And battery capacity diminishes with age and recharges.

If an electric car runs out of power on a highway before it can find a charging station to recharge, it is a problem. Gasoline fueled cars can easily get a can of gas. An electric car would have to be towed to a charging station at considerable expense, and disrupt normal traffic flow.

There are on the order of 250,000 gasoline stations in America. It is likely that there will be a far smaller number of charging stations, making them harder to find and access. And charging times will be much longer. It only takes 2 or 3 minutes to fill one's gas tank, but even so, there often are multiple cars at a gas station at the same time. With charging times of 20 or 30 minutes and fewer charging stations, there will be many more cars waiting to get charged.

Finally, electric cars are more expensive than gasoline fueled cars, a burden for lower income families and individuals.

Maglev Networks will enable fast, very low cost non-highway, long distance travel for passengers, trucks, and autos. For long distance trips, electric cars and trucks could travel on Maglev vehicles, instead of needing a very large battery pack, or charging at multiple stations on the journey. Plus, they could be electrically charged while on the Maglev car-carrying vehicle. Electric cars and trucks could then operate locally with smaller, less expensive battery packs and more convenient charging locations.

Second, the Maglev Public Transit systems will substantially reduce the need for electric cars. Today, about 75% of people going to work get there in their own cars, because the public transit systems are slow, crowded, expensive and inconvenient, with poor service frequency. With Maglev Public transit, transport will be much more attractive, with a much greater percentage of workers using it rather than driving to work.

For locations where Maglev service is not available, motor vehicle transport will be necessary, either electric, or gasoline/diesel fueled. For the fueled vehicles the fuel would not have to use fossil fuels, however. Instead, as described in Chapter 7, using low cost beamed solar power from space, synthetic gasoline/diesel fuel can be manufactured from carbon dioxide extracted from the atmosphere and hydrogen generated by electrolyzing water. The cost per gallon of synthetic fuel will be comparable to the present cost of gasoline/diesel fuel from fossil oil.

While electric cars will be significant in America's transport future, High Speed Rail (HSR) will not. At the present rate of about 15,000 passenger miles per capita, during an 80-year lifespan, Americans travel the equivalent of 1.2 million miles, 48 times around our 25,000-mile circumference World. Traveling on intercity rail (Amtrak) at the present, an average of 20 miles per capita per year, would be equivalent to 1,600 miles, one round trip between Chicago and New York City, during a person's 80-year lifespan.

Building a few HSR corridors would increase intercity passenger rail traffic, but it would still be insignificant. Today, 8 million passengers per year fly between San Francisco and Los Angeles, while 3 million fly between Boston and Washington, DC. If they all were to switch from air to HSR, a very unlikely scenario, intercity rail passenger traffic would increase by 11 million passengers per year, from the present 26 million passengers to 37 million. The average rail travel for Americans would increase to about 30 miles per capita per year. Not very impressive! One round trip between New York City and Miami, Florida in an 80 year lifetime.

The HSR construction expense would be enormous. A projected (before cost overruns) of $150 Billion for a Boston to Washington, DC line and $70 Billion for San Francisco to Los Angeles. Many Billions more in operating subsidies over the years. Almost every HSR route in the World requires substantial government subsidies.

And the high cost per passenger mile. Riders on the Acela between NY City and Washington, DC pay $0.75 per passenger mile. At that rate, a San Francisco – Los Angeles round trip would cost $600. Not many could afford it – just the top 1 percent – the 99 percent would have to travel some other way. But, they would still have to pay taxes to subsidize the HSR line that carried the 1%.

Simple math shows how useless HSR would be for America. We are a very big country in size, with a very low population density compared to countries that have HSR routes. Japan has 337 persons per square kilometer, a factor of 10 higher than the United States 32 persons per square kilometer, Germany has 229 persons per square kilometer, a factor of 7 higher. China has 142 persons/square kilometer, a factor of 4 higher. Even with countries with much higher population densities, HSR accounts for only a small percentage of total passenger travel – e.g., in France, average per capita HSR travel distance is 400 miles, compared to 7,600 miles on the highway.

In contrast to biofuels, hydrogen cars, and High Speed Rail lines, 2nd Generation Maglev is a practical path forward for America's and the World's future transport needs. Together with electric cars and synthetic hydrocarbon fuels, 2nd Generation Maglev will provide faster, better, and safer transport at lower cost than today's transport systems, without having to consume vast quantities of fossil fuels that are very destructive to Earths environment and human health.

In the remainder of the chapter we describe potential 2nd Generation Systems for the US and the World – where they could be built, their transport capabilities, economic and environmental benefits, the cost of construction, and how soon they could be implemented. We first describe 2nd Generation Maglev Systems for the US, and then 2nd Generation Maglev Systems for the World.

## 2nd Generation Maglev for the United States – Intercity and Public Transit Systems

Today, US transport is a mess, and it will get much worse. In the years ahead, if we continue on our present path. The realities are seen in the list of DOT statistics given below.

We spend an enormous amount on transporting people and goods – 1.5 Trillion dollars per year, 10% of US Gross Domestic Product, $8,300 per household, as much as we spend on food plus clothing.

On average, Americans travel about 15,000 miles per year, more than halfway around the World. Sadly, it's not "See the World" travel. 88 percent (12,600 miles) is on congested bumpy highways with lots of potholes, or jammed together in noisy public transit buses, subways, and commuter rail cars. 12 percent (1,730 miles), on crowded airplanes that are often late. And travel on slow, bumpy intercity trains? 20 miles per year per person.

And the highways are very dangerous – 33,000 persons died on America's highways in 2010, with 3 million injured. The medical, insurance, health damage, and long term quality of life cost? $877 Billion per year, according to the National Highway Traffic Safety Administration. [*The Economic and Societal Impact of Motor Vehicle Crashes, 2010 NYTSA DOT HS 812013*]

Americans own 230 million cars, with an average of 0.83 cars per person in our population of 312 million. On average, each American travels 11,500 miles per year on our highways. Today, congestion delays are estimated to cost the US $100 Billion dollars per year. The DOT projects that highway congestion will increase by 366 percent in 2040 AD. In 2035 on the 1,381 mile I-5 highway from San Diego to the Canadian border north of Seattle, 95% of the 550 miles of urban segments will be congested, with 85% of the rural segments congested. Traffic flow on the I-5 Highway will be enormous, with a maximum of 600,000 vehicles and 70,000 trucks per day.

Average Daily Long-Haul Freight Traffic on the National Highway System 2002

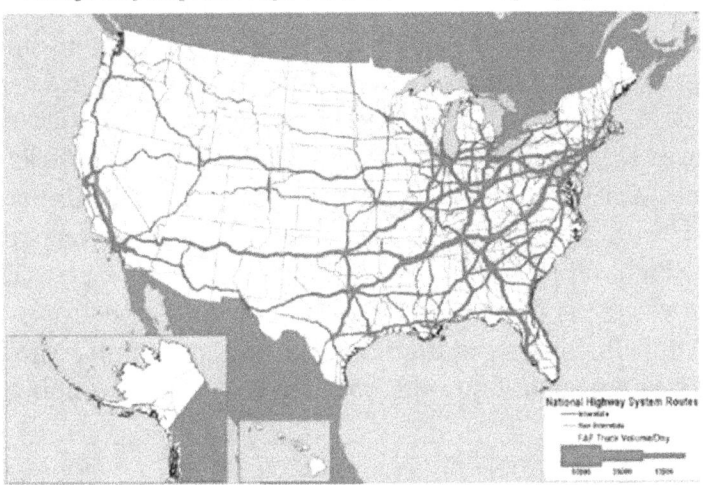

Average Daily Long-Haul Freight Traffic on the National Highway System 2035

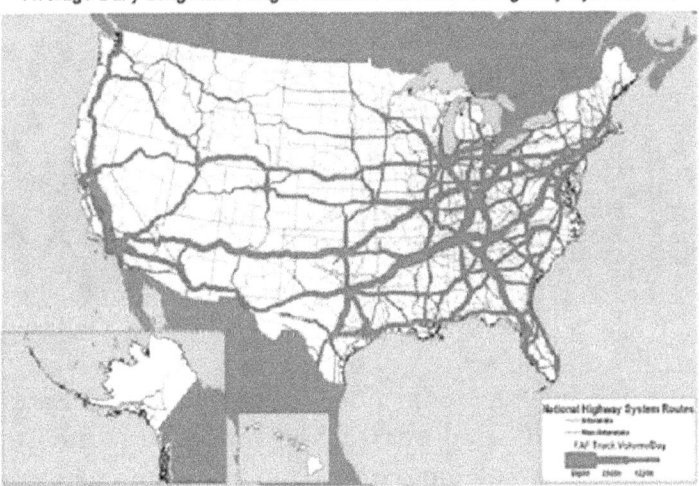

Figure 4.6

Highway trucks are a vital part of America's transport network. In 2011, highway trucks moved 11 Billion tons of goods (35 tons per capita) worth 10.5 Trillion dollars (66% of US GDP), at a cost of 500 Billion dollars annually for truck operations. And truck transport will almost double by 2040, with projected movement of 17 Billion tons worth 21 Trillion dollars. Figure 4.6 compares the present US truck traffic flow with truck traffic flow in 2035.

Freight rail remains the cheapest in cents per ton-mile, but present revenues for highway freight trucks are much greater than for freight rail, 700 Billion dollars per year for truck freight compared to 70 Billion for rail freight. The reason why truck freight revenues are much greater? Trucks deliver their loads in much shorter times and pickup/deliver loads much more conveniently. Today, high value freight goes by highway truck – 10.5 Trillion dollars' worth per year – while low value freight – coal, iron ore goes slowly by rail. Tomorrow, intercity highway trucks will go 300 mph on Maglev, instead of 50 mph on highways, at considerably lower costs.

Figure 4.7 Accident on US Highway

The safety and environmental benefits of Maglev? Enormous! The National Maglev Network will save many 10's of thousands of lives and hundreds of thousands of injuries now happening in accidents (Figure 4.7) on America's highways every year. It will save hundreds of Billions of dollars annually now lost to highway accidents. It will greatly reduce the 5 Billion barrels of fossil fuel we now use for transport and the 1.8 Billion tons of $CO_2$ greenhouse gas emitted from our tailpipes and jet engines. It will also greatly reduce the damage to our hearts and lungs from the pollutants and microparticles emitted by our 230 million automobiles and 10 million trucks.

The benefits to the economy and our quality of life will be tremendous. By taking trucks and autos off the road, not only will the National Maglev Network substantially reduce highway deaths and injuries, it also will greatly reduce highway traffic congestion (Figure 4.8) and delays, which today cost the US economy 100 Billion dollars annually. This, plus reducing the 900 Billion dollars now spent on medical expenses, insurance, health damage and lost income from highway accidents, plus Maglev's considerably lower cost per passenger mile and truck ton mile, will greatly benefit the US economy, saving each of us more than $1,000 per year.

Figure 4.8 Congestion on a US Highway

Figure 4.9 shows a map of America's completed 29,000-mile National Maglev Network, built on rights-of-way of the US Interstate Highway System. It serves all lower 48 continental US states, plus Toronto, Montreal, and Vancouver in Canada. It connects all metropolitan areas in the US with populations of 250,000 persons or more. 70 percent of the US population would live within 15 miles of a Maglev station, from which they could travel to any other stations in the Network at 300 mph average trip speed.

The Network would transport passengers, passengers with autos, and highway trucks, at lower cost and much faster than highway driving. Passenger trip times would be comparable to those for air travel at lower cost, with much more frequent and comfortable service.

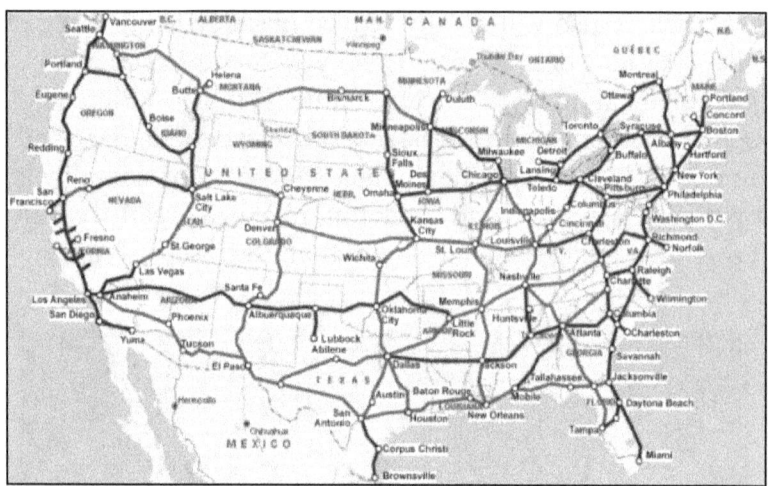

Figure 4.9
Map of the US National Maglev Network

Table 4.0:  Population and States Served by US National Maglev Network

| Maglev Network | States In Network | Population of States in Network (millions) | Population Living Within 15 Miles of Stations (millions) | Route Miles in Network |
|---|---|---|---|---|
| First, Second and Third Waves Completed | 48 plus Toronto, Montreal & Vancouver | 315 includes Toronto, Montreal & Vancouver | 232 includes Toronto, Montreal & Vancouver | 29,000 |
| 74% of population in States live within 15 Miles of a Station | | | | |

Following a 5-year development period for the 2nd Generation Maglev System, the 29,000-mile National Maglev Network would be built in 3 phases, or "waves", over a 15-year period, with each wave lasting 5 years. The First Wave is shown in Figure 4.10 and Table 4.1. It would be constructed on the US East and West Coasts where most Americans live.

Figure 4.10: First Maglev Wave to Be Completed 10 years from Start of US Maglev Program

Table 4.1 Population and States Served in First Wave

| Maglev Network | States In Network | Population of States in Network (millions) | Population Living Within 15 Miles of Maglev Stations (millions) | Route Miles in Network |
|---|---|---|---|---|
| East Coast/Midwest Network | 45<br>MN, WI, IL, IN, OH, PA, NY, MA, VT, NH, MN, ME, RI, DE, MD, VA, DC, NC, SC, GA, FL +Toronto & Montreal | 175.8<br>(includes Toronto, Montreal) | 102.9<br>(includes Toronto, Montreal) | 4,224 |
| West Coast Maglev Network | CA, NV, OR, WA & Vancouver, Canada | 50.9<br>(includes Vancouver) | 43.5<br>(includes Vancouver) | 2006 |
| Total for First Maglev Wave (Both Networks) | 26 States Plus Toronto, Montreal & Vancouver | 226.7 | 146.4 | 6230 |

65 % of population in States Served by the Networks live within 15 Miles of a Maglev Station

Figure 4.11 shows a more detailed map of the West Coast portion of the First Wave. It would connect 51 million persons to the West Coast Maglev System – 36.5 million in California, 3.8 million in Oregon, 6.5 million in Washington State, 2.6 million in Nevada, and 1.3 million in Vancouver, British Columbia. 85% of the population would live within 15 miles of a Maglev station. The San Diego to Vancouver route along Interstate 5 would be 1,380 miles in length. Side Maglev routes along I-15 to Nevada plus side routes in California would increase the West Coast Systems to a total of 2,000 route Miles.

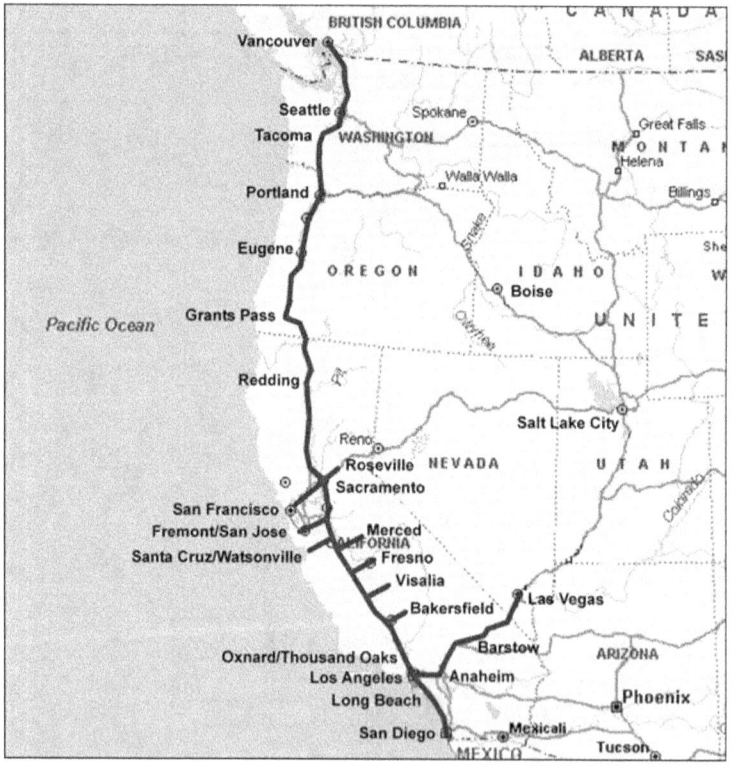

Figure 4.11 Map of the West Coast Maglev Network

Table 4.2
Compares trip times for Maglev to trip times by air, rail and highway.

| Illustrative Trip | West Coast Maglev Passenger, Auto , or Truck | Air [1] Passenger | Rail Passengers [2] | | Highway, Auto & Truck |
| --- | --- | --- | --- | --- | --- |
| | | | Conventional Rail | High Speed Rail | |
| San Diego to Seattle | 4 Hrs 30 min | 4 Hrs | 21 Hrs | 9 Hrs 40 min | 25 Hrs 15 min |
| San Francisco to Los Angeles | 1 Hr 45 min | 2 Hrs 30 min | 6 Hrs | 3 Hrs 45 min | 9 Hrs 40 min |
| Portland to San Francisco | 2 Hrs 30 min | 2 Hrs 45 min | 10 Hrs 45 min | 4 Hrs 50 min | 12 Hrs 45 min |
| Los Angeles to Las Vegas | 1 Hr | 2 Hrs | 4 Hrs 40 min | 2 Hrs 10 min | 5 Hrs 30 min |

Includes 1 Hr Pre-Boarding Time at Airport for Check-In
Average Speed of 60 mph by Conventional Trains (Amtrak)
Average speed of 130 mph by High Speed Rail (French TGV)
Average Highway Speed of 50 mph (Including Congestion Delays & Rest Stops)

Table 4.2 compares trip times for Maglev, air and highway travel for 4 illustrative trips – San Diego to San Francisco, and Los Angeles to Las Vegas. Maglev has the shortest trip time except for air travel between San Diego and Seattle. However, Maglev is very competitive with air travel for those trips, and will be even faster when airplane boarding and check-in times are longer than 1 hour. Taking into account the very frequent Maglev service – along with flight delays and the lower frequency of airline service, plus lower fares on Maglev, it appears very likely that Maglev would be the choice of travelers. Table 4.3 shows the projected cost using Maglev transport for the 4 illustrative trips on the West Coast Maglev System based on the assumed Maglev traffic flows shown in Table 4.3. The costs are substantially lower than by air, rail, or highway.

The West Coast Maglev System will substantially reduce highway congestion along the I-5 Corridor. In 2035, traffic flow is projected by the US Department of Transportation to increase to 2 times the flow in 2007. (Table 4.4) Without Maglev, most urban plus rural segments of I-5, will be highly congested, greatly increasing trip times and congestion costs.

Taking the actual 2007 traffic flow in the I-5 corridor for comparison, the assumed Maglev traffic flow along the I-5 corridor in the future West Coast Maglev Network (Table 4.3) corresponds to:

- 30,000 passengers daily along the I-5 Corridor in the future West Coast Maglev Network (100 per maglev vehicle) (42% of auto/passenger traffic on 2007 I-5)
- 5,000 trucks daily (50% of 10,000 truck traffic on 2007 I-5)
- 20,000 person autos w/passengers daily (10 per vehicle) (71,000 28% of auto traffic on 2007 I-5)

2035 total traffic flow on the I-5 Corridor will be approximately twice that in 2007, so Maglev flow on the I-5 Corridor in 2035 will likely be considerably greater than that shown in Table 4.3, when the National Network is in full operation.

Using the unit costs for Maglev transport of passengers, highway trucks and autos (cents per passenger mile, cents per auto mile, and cents per ton mile) shown in Table 4.5, compared with existing transport costs, and assuming the above Maglev transport volumes, annual savings would be 21 Billion dollars (Table 4.6) in terms of the 2007 AD traffic flow. The projected savings, of course, cannot occur, because it now is 2015, not 2007. When the West Coast Maglev System is operating, with the 2035 projected traffic (Table 4.4), 2 times that in 2007, the Maglev annual transport savings would be 42 Billion dollars (2x21 Billion/year).

The West Coast Maglev Network is extremely attractive in terms of faster trip times, lower costs, increased safety, greater energy efficiency, and decreased greenhouse gas emissions. A more detailed description of the West Coast Network is given in "Maglev America", available at Amazon.com.

Table 4.3

| Trip Times & Costs on the West Coast Maglev Network | | | | | |
|---|---|---|---|---|---|
| Illustrative Trip | Trip Miles | Trip Time | One Way Maglev Trip Cost | | |
| | | | Passenger | Auto w/ Passengers | Highway Truck * |
| San Diego to Seattle | 1260 | 4 Hrs 30 Min | $37.80 | $403 | $128 |
| San Francisco to Los Angeles | 480 | 1 Hrs 45 Min | $14.40 | $154 | $49 |
| Portland to San Francisco | 640 | 2 Hrs 30 Min | $19.20 | $205 | $65 |
| Los Angeles to Las Vegas | 275 | 1 Hr | $8.00 | $88 | $28 |

Basis: 280 mph Avg Speed, Daily Avg Traffic 30,000 Passengers, 20,000 Autos w/Passengers, 5,000 Highway Trucks, 10 Cents/KWh, Unit Capital Cost, 30 M$/2-Way Mile for Guideway, 5M$/Maglev Vehicle, *$ Per Ton of Load, 30 Tons Load per Truck

Table 4.4
Vehicle Flows and Congestion Along the I-5 Corridor

| | Traffic Flow on Corridor | | | |
|---|---|---|---|---|
| | 2007 | | 2035 | |
| | Avg | Max | Avg | Max |
| Vehicles/Day | 71,000 | 300,000 | 150,000 | ~600,000 |
| Trucks/Day | 10,000 | 35,000 | 22,000 | ~70,000 |
| Urban Segments,* % Congestion | 65% | | 95% | |
| Rural Segments, % Congestion | 31% | | 85% | |

*(550 miles of 1381 mile total length are urban segments)

Table 4.5

Comparison of Maglev Travel Costs with Costs for other Modes

|  | Unit Cost | Maglev [1] | Highway [2] | Air [3] | High Speed Rail [4] |
|---|---|---|---|---|---|
| Passengers | Cents/passenger mile | 3 | --- | 15 | 50 |
| Autos | Cents/mile | 32 | 40 | --- | --- |
| Trucks | Cents/ton mile | 10.2 | 30 | --- | --- |
| Seattle to San Diego Trip (1,260 miles one way) Dollars | | | | | |
| Passengers | | $37.80 | Travel w/Auto | $190 | $630 |
| Auto w/Passengers | | $403 | $500 | --- | --- |
| Trucks | | $128 | $480 | --- | --- |

1) Costs include amortization and maintenance of guideway & vehicles

2) Avg cost of operating a car (depreciation, fuel, tires, etc)

3) Avg cost per passenger mile (US Statistical Abstract)

4) Cost/passenger mile for High Speed Rail in Europe


Table 4.6

Annual Savings in Transport Cost Assuming 2007 Traffic Flow Enabled by West Coast Maglev Network

|  | Maglev Travel (B$ per year) | Existing Mode of Travel (B$ per year) |
|---|---|---|
| Passengers | 0.64 B$ | 6.6 B$ (Highway & Air) |
| Autos w/Passengers | 4.7 B$ | 5.8 B$ (Highway) |
| Highway Trucks | 7 B$ | 21 B$ |
| Total | 12.3 B$ | 33.4 B$ |
| Savings by Maglev = 21B$ per year | | |

Turning now to the East Coast Maglev Network, the other part of the First Wave of the National Maglev Network, Figure 4.12 shows a map of the Northeast-Midwest Section of the East Coast Maglev Network, down to Richmond, Virginia. The Richmond to Miami section is shown in Figure 4.10, the map of the First Wave.

While the Chicago to Albany, New York Maglev route is not geographically part of America's East Coast, it is very important to include it in the Maglev East Coast Network, because of the enormous truck traffic from the Chicago region to New York and other cities on the East Coast.

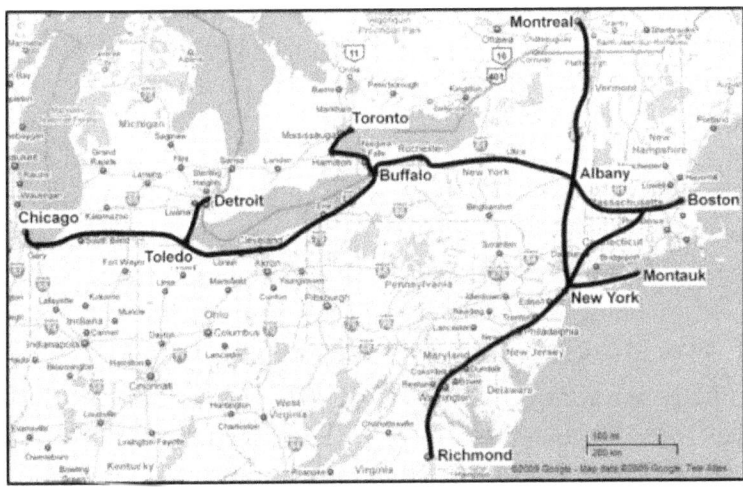

Figure 4.12
Map of the Northeast-Midwest Portion of the East Coast Maglev Network

Summarizing, the West Coast Maglev Network has 2,108 miles of guideway, costing 63 Billion dollars. 43 million Americans and Canadians live within 15 miles of a Maglev Station on the West Coast Maglev Network. The East Coast Maglev Network has 4,224 miles of Maglev guideway. Cost is 126 Billion dollars. 103 million Americans and Canadians live within 15 miles of a Maglev Station on the East Coast Network. Total population living within 15 miles of a Maglev Station for the First Wave of the National Maglev Network, to be completed in 10 years from Start, would be 146 million.

It's interesting to compare the First Wave with the proposed High Speed Rail lines from Boston to Washington (estimated cost of 150 Billion dollars) and San Francisco to Los Angeles (estimated cost of 70 Billion). Cost for the First Maglev Wave? Comparable. 190 Billion dollars. Population served by the HSR lines? About 20 million riders. Population served by the First Maglev Wave? 146 million. Passenger fare by HSR, about $0.75 per passenger mile. Passenger fare by Maglev? About 10 cents per passenger mile, enabling a much larger ridership.

Figure 4.13 shows a map of the Second Wave of the National Maglev Network, to be built from year 11 of the National Maglev Network project through year 15. It would connect America's West and East Coast Maglev Networks built in the First Wave, with 3 Transcontinental Maglev routes, located in the Northern part of the US, the Middle part of the US, and the Southern part along the Gulf Coast. Additional routes would be built to connect inland cities to the Second Wave network. Total route mileage would increase from the 6,230 miles in the First Wave to 18,630 miles when the Second wave was completed (Table 4.7). The population living within 15 miles from a Maglev station would grow from the 146 million in the First Wave to 210 million in the Second Wave.

Figure 4.14 shows a map of the Third Wave of the National Maglev Network to be built from Year 16 through Year 20 of the 20 year long project. North-South Maglev routes would be built to connect America's metro/micropolitan areas into an even closer knit Network, with shorter trips between the North and South of the US. Total route mileage would increase from the 18,630 miles the Second Wave to 29,000 miles when the Third Wave was completed. The population living within 15 miles of a Maglev station would increase from 210 million to 230 million (Table 4.8).

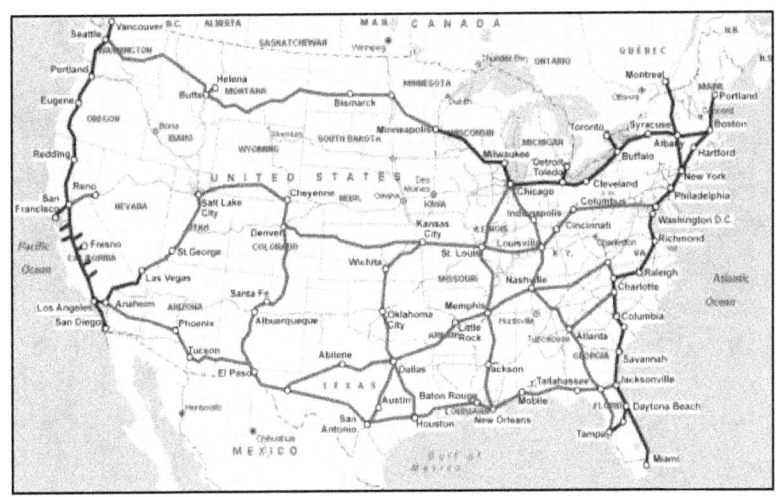

Figure 4.13
Map of the Second Wave of the US National Maglev Network

Table 4.7:  Population and States served in Second Wave

| Maglev Network | States in Network | Population of States in Network (millions) | Population Living Within 15 Miles of Maglev Stations (millions) | Route Miles in Network |
|---|---|---|---|---|
| First Wave Plus Second Wave | 45 (Iowa, Nebraska & S. Dakota not in Network) plus Toronto, Montreal & Vancouver | 310 (includes Toronto, Montreal & Vancouver) | 210 (includes Toronto, Montreal & Vancouver) | 18,630 |

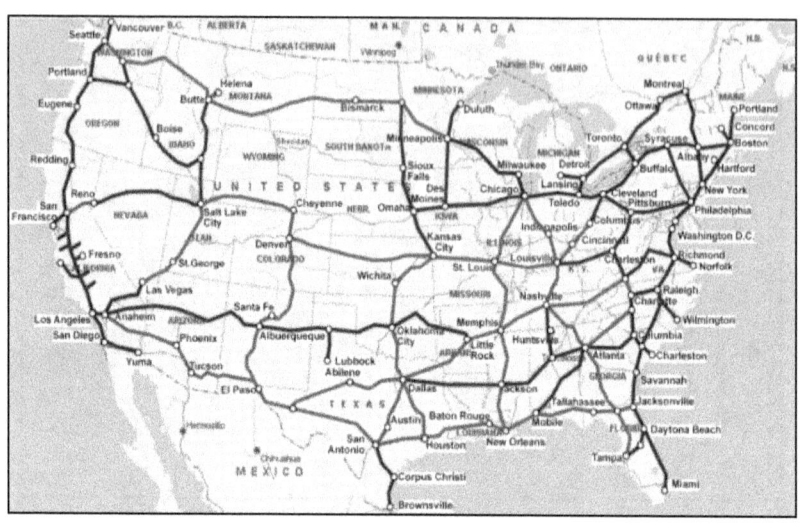

Figure 4.14: Map of Third Wave of National Maglev Network

Table 4.8: Population and States Served in Third Wave

| Maglev Network | States In Network | Population of States in Network (millions) | Population Living within 15 Miles of Stations (millions) | Route Miles in Network |
|---|---|---|---|---|
| First, Second and Third Waves Completed | 48 plus Toronto, Montreal & Vancouver | 315 includes Toronto, Montreal & Vancouver | 232 includes Toronto, Montreal & Vancouver | 29,000 |

74% of population in States live within 15 Miles of a Station

## Maglev Public Transit

Working in tandem with the National Maglev Network, there will be many Maglev Public Transit Systems that operate in America's metropolitan areas. The 2nd Generation Maglev-2000 System not only is the basis for the National Maglev Network, but also can provide faster, better, cheaper Public Transit.

The 2nd Gen Maglev System has the unique capability to be adapted at very low cost to existing railroad and subway tracks for travel by magnetic levitated and propelled passenger vehicles. The adaptation can be done at night and low travel periods, without interfering with conventional train and subway schedules, during times when travel by the existing equipment is infrequent. The adaptation for Maglev travel is easy and quick. Simple panels containing loops of ordinary aluminum conductor wire are attached to the crossties of the railroad and subway tracks. Conventional trains and subway cars can continue to use the tracks after they have been adapted for Maglev, if desired.

## The Benefits of Maglev Public Transit?

Lower operating costs – much less maintenance required for tracks and vehicles, more energy efficient, increased employee efficiency and productivity, more convenient and more frequent service, shorter trip times, much lower government subsidy requirement, low fares, and much more comfortable and healthier travel.

As an example, the Long Island Railroad (LIRR) System, is the largest commuter rail system in the United States. It carries 280,000 passengers per day on weekdays, with a total of 81 million passengers per year – 3 times the total Amtrak ridership for all America. The average LIRR fare cost paid by passengers is 26 cents per passenger mile, while the actual average cost per passenger mile is 80 cents, with the difference of 54 cents per passenger mile paid by government subsides. With the Maglev LIRR, the government subsidy will be much less.

For the whole LIRR System, the average passenger fare is $6.46, 45% of the $14.68 actual cost. The difference is government subsidies. Adaptation of the LIRR for Maglev service would result in major benefits to passengers, taxpayers, and people living near the LIRR tracks, including:

- Much lower taxpayer subsidies
- Much shorter trip times, a factor of 2 to 3 shorter using higher speed, faster accelerating Maglev vehicles.
- Lower passenger fares
- Very quiet operation, no rail or locomotive noise for passengers and people living near the LIRR tracks
- More frequent service – individual Maglev vehicles, no infrequent long trains of many cars pulled by W
- More comfortable rides: no vibration, bumping and swaying of RR cars, less crowded passenger seating
- No diesel emissions of greenhouse gases and microparticulates
- Safer operation – no 3rd rail, able to stop much faster in emergencies

Cost and schedule for adapting the LIRR to Maglev? For 700 miles of one-way track, the capital cost at 4 million dollars per one-way mile would be 2.8 Billion dollars, about 93 million dollars annually over a 30 year amortization period. The track adaptation annual cost would be approximately 5% of the annual LIRR budget. Put in another way, the 2.8 Billion dollars to adapt 700 miles of LIRR track is about 1/4 of the 10 plus Billion dollars the LIRR is now spending to dig a tunnel under the East River to connect the LIRR to Grand Central Station in New York City.

At 5 million dollars per Maglev vehicle, the cost of 300 vehicles to transport the LIRR's 280.000 daily riders would be approximately 1.5 Billion dollars, about 1/2 of the track adaptation cost. With mass production of Maglev vehicles, the unit cost will probably be less than 5 million dollars.

With a Maglev LIRR, trip times will be much shorter. The average speed of LIRR trains is about 30 mph – a result of the slow acceleration and deceleration for conventional long trains of many cars, and the requirement that the train stop at many stations along its route. Maglev LIRR vehicles will travel as individual units, able to accelerate and decelerate much faster, like ordinary automobiles, and able to travel past stations at full speed that they do not have passengers for.

Riders on the Maglev LIRR will love it. Trip times a factor of 2 shorter. Babylon to Montauk, a distance of 79 miles. Today's travel time is 2 hours 22 minutes an average of 33 mph. On Maglev LIRR, it would be 1 hour 11 minutes, an average of 66 mph. There are presently 6 long trains per day on the Babylon – Montauk Branch. With Maglev LIRR, it could be 20 or more vehicles per day for the trip, much more convenient service. And, no more noisy, bumpy, and swaying rides. Just quiet, comfortable, no vibration – like sitting in a chair in the living room.

Maglev can be adapted to other commuter rail systems in the US, like Metro North in New York State. We have considered adapting Maglev to US light rail systems; however, light rail ridership generally appears too low to be cost effective, and adaptation would be more difficult and expensive than for heavy rail and commuter rail.

2nd Generation Maglev can also be adapted to existing subway systems, in particular the New York City Subway System. Figure 4.15 shows a New York City Subway car. The NYC Subway System is a marvel. It transports 6.5 million passengers daily. NYC's annual ridership is 2.4 Billion, 1/4th of the 10.4 Billion total annual US transit ridership for all modes – commuter rail, subways, and buses. However, as anyone who has ridden the NYC subway knows, it is not the most pleasant ride. Noise levels are astronomic, reaching 100 decibels at some stations, with possible hearing damage. Riders are jammed together in very crowded, bumping and swaying cars, breathing in steel dust and other particulates from erosion of rails and brakes.

Figure 4.15 New York City Subway Car

Adaptation of the NYC Subway, and other transit systems in the US for Maglev will provide much better ride quality – no noise, no bumping and swaying of the transit cars, less crowded, more frequent service, and much cleaner air – no brake or rail dust to breathe in. As with the Maglev LIRR, operations will be cheaper and more efficient, and maintenance will be much less, enabling substantial reductions in government subsidies for public transit.

Total US annual ridership on all public transit modes – heavy rail, commuter rail, light rail, and buses – is 10.4 Billion passengers. 30% of the total ridership, 3.5 Billion passengers travel on heavy rail systems –NYC Subway, Washington DC Metro, San Francisco Bart, and others. Of the 3.5 Billion heavy rail passengers, 2.4 Billion (70%) use the NYC Subway system.

Figure 4.16 shows a map of the NYC Subway System. In terms of government subsidies, it performs very well compared to other public transit systems. The average passenger fare per trip is $1.05, with an actual operating cost per trip of $1.40. The average fare per passenger trip for all US transit modes is $1.18, while the actual cost is $3.54 per trip, 3 times the fare cost. The $2.26 subsidy per trip is paid by taxpayers.

Adapting the NYC Subway System to Maglev will result in many benefits:

- Reduced subsidies from taxpayers
- Faster, much more comfortable trips – no bumping and swaying
- Quiet trips – no 100 decibels noise, which causes hearing loss
- No breathing in steel dust and other health harming microparticles generated by braking on steel rails
- Greater energy efficiency
- Similar benefits will result from adaptation of Maglev to the other US heavy rail systems.

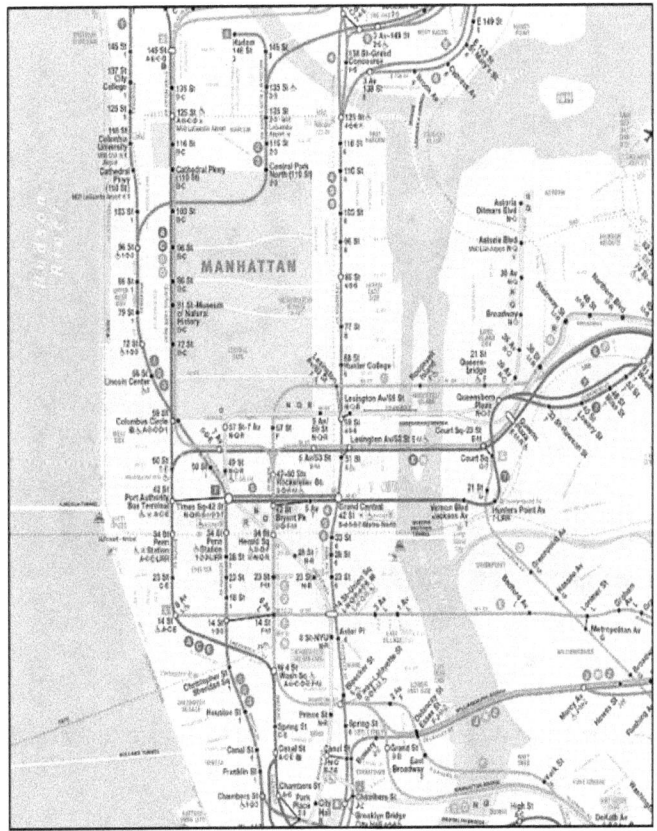

Figure: 4.16 NYC Subway System

Details of the adaptation process for the NYC Subway System for Maglev operation is described in *"Maglev America"*. Summarizing, the capital cost of the installation of the aluminum loop panels on the cross ties of the subway track (Figure 4.17) plus the capital cost of the Maglev vehicles and their superconducting Magnets is projected to be 10 Billion dollars. Amortized over 30 years that would be 330 million dollars per year, 10 percent of the NYC subways present operating budget of 3.3 Billion dollars per year. The savings in operating costs made possible with Maglev would more than offset the adaptation cost. The adaptation process could be carried out in as little as 2 years, given adequate funding and high priority for the program.

Figure 4.17
New York City Subway Track Adapted for Maglev Service

In summary, the National Maglev Network and Maglev Public Transit will be of great benefit to America in its capability for:

- Much lower cost of transport
- Faster and more comfortable travel with shorter trip times
- Safer, less congested highways, with substantial reductions in deaths and injuries
- Greater energy efficiency and reduced pollution
- Increased economic productivity.

## 2nd Generation Maglev Systems for the World – The Global Maglev Network

The Global Maglev Network – Why it is Needed. By 2050 AD, only 35 years from now, World leaders and experts project that the World economy and its population will grow greatly. Compared to now (2014), by 2050:

- World population will grow from today's 7 Billion to 9 Billion. GDP will almost triple from approximately 70 Trillion US$ annually to more than 190 Trillion US$.
- Electricity generation will double, from 20 Trillion Kilowatt Hours per year, to 39 Trillion Kilowatt Hours.
- Number of automobiles will increase from 1 Billion to 2.5 Billion, a factor of 2.5
- Passenger miles will increase by a factor of 2.6
- Tonne miles of freight will increase by a factor of 2
- Global carbon dioxide emissions from transport will increase by a factor of 2.1, assuming that transport continues to be based on fossil fuels.

The World economy and population will not be sustainable at these levels if we continue to depend on fossil fuels for energy. As described earlier in chapters 2 and 3, global warming, ocean acidification, and depletion of fossil fuel reserves will lead to the collapse of modern civilization if humanity continues to use fossil fuels as its primary source of energy.

Transport of passengers and freight is critically important to modern civilization, today, and will continue to be in the future. However, transport must transition from fossil fuels to a different and sustainable source of energy, and must do soon. Electrification of transport, i.e., electric cars and rail, together with reductions in air travel and water-borne shipping, which cannot be electrified, is the only solution.

Maglev will make the electrification of global transport practical in 3 ways. First, by transporting passengers and freight, at high speed, high efficiency and low-cost, on Maglev guideways built between metropolitan areas, greatly reducing intercity travel by highway and airplanes. Second, by adapting already existing railroad and subway tracks at very low-cost for Maglev public transit, in urban and suburban areas. Maglev public transit will be much faster, cheaper, more comfortable, more frequent in service, and better environmentally. Subsidies from taxpayers will be much less. These advantages will help shift travel by highway vehicles to travel on public transit systems.

Third, with an intercontinental global Maglev Network, transport of freight by container and other ships will be much less, greatly reducing oil consumption and greenhouse gas emissions from shipping. It takes 35 days to ship containers from Asia by water. On Maglev it will be only a day, and shipping cost will be less.

For Maglev to be a major mode of transport in the United States, it must function as a National Network. Building isolated routes between a few city pairs, like the proposed Los Angeles to San Francisco High Speed Rail (HSR) System or the Boston-New York City, Washington, DC HSR route, will do very little for US transport needs.

To be effective, transport systems must be a connected network that serves the whole country like the US Highway system, the US railroad system, and the US airplane/airport system. Isolated routes between 2 points, with no connection to other points in the country, would be of little use.

Similarly, to be greatly effective globally, Maglev must be a Global Maglev Network. Described in more detail later, here we list its highlights. The Global Maglev Network will:

- Interconnect the World's principal continents – Asia, Africa, Europe, North America and South America.
- Connect to the principal cities in each continent
- Adapt existing public transit rail systems for Maglev service where appropriate
- Transport passengers, trucks, and autos on the same guideway or railroad trackage adapted for Maglev
- Transport a major portion of the containers and freight presently transported by ship.

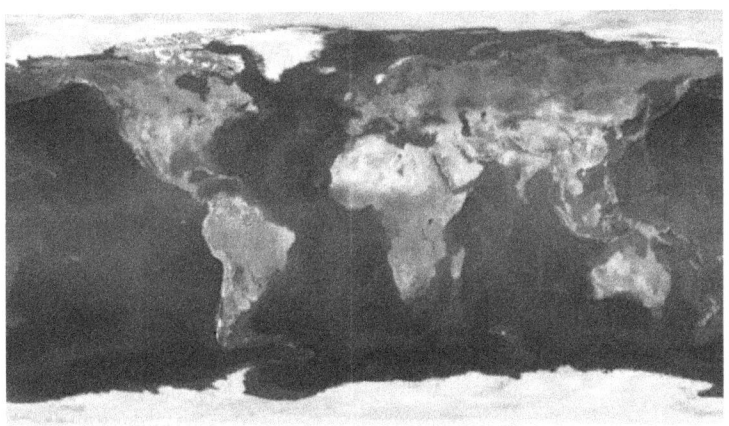

Figure 4.18 Map of the World

The World is very, very big (Figure 4.18) 25,000 miles for a trip around the World, 197 million square miles in area, 3/4th of it water. Very high mountains, immense oceans, very large deserts – all major challenges for creating a Global Maglev Network. The US National Maglev Network described earlier has 29,000 miles of high speed Maglev intercity routes. The Global Maglev Network will require many more miles.

Starting with the Maglev intercontinental connections, the Europe to Africa Maglev connections would be made through the

Strait of Gibraltar (Figure 4.19) using either a tunnel underneath the seabed, similar to the English Chunnel from Britain to France, or a floating undersea tube anchored to the sea floor. The tube could be at a sufficient depth, on the order of 100 meters, that wave movements would be very small. Using an undersea tube, the crossing length would be only 14 miles, much shorter than the 31 mile long Chunnel.

Figure 4.19  The Strait of Gibraltar

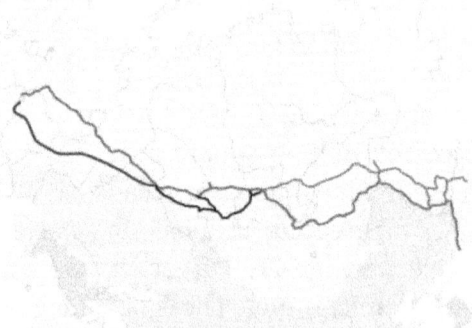

Figure 4.20  The Trans-Siberian Railway

Europe and Asia are actually one continent with the distinction being cultural, not geographical. The great distance between the societies in Europe and those in the Far East, e.g., China and Japan, makes them seem like separate continents. Historically, there has been land transport of humans and goods between Europe and Asia for centuries – Marco Polo, the Silk Road, Genghis Khan, and so on.

The Europe to Asia Maglev Connection will in general follow the same path as the existing Trans-Siberian Railway that runs 5,753 miles across Russia from Moscow to Vladivostok on the Pacific Ocean (Figure 4.20). It takes 6 days from Moscow to Vladivostok and Beijing on the Trans-Siberian Railway. By 300 mph Maglev, it will take only 1 day. It takes 35 days by container ship from China to Europe. With Maglev it will be only 1 day, and it will be cheaper.

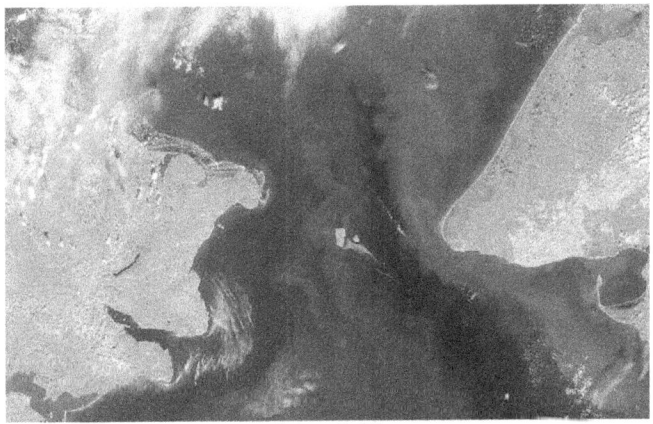

Figure4. 21
The Bering Strait Separating Siberia from Alaska in the
North Pacific. NASA image taken by MISR satellite

The Asia to North America Maglev connection would be made across the Bering Strait (Figure 4.21) through either a tunnel beneath the seabed or a tube resting on the sea floor. The concept of a bridge or tunnel across the Bering Strait is over 120 years old, and has been proposed many times over the years. The Bering Strait crossing would connect Russia's Chukchi Peninsula to America's Diomedes Islands, located midway in the 50 mile crossing, and then from the Diomedes Islands to Alaska's Seward Peninsula.

A Maglev System would then connect Alaska through Canada to the US National Maglev Network. From the US, Maglev routes would run South through Mexico, Central America and Panama, connecting to Maglev routes along the West Coast of South America to form the Pan American Maglev Highway (Figure 4.22), extending all the way to the bottom of South America. The Pan American Maglev Highway would directly serve Columbia, Ecuador, Peru, and Chile. Additional Maglev Systems (not shown) would connect the other countries in South America, i.e., Venezuela, Brazil, Argentina, Bolivia, Paraguay and Uruguay to the Pan American Maglev Highway.

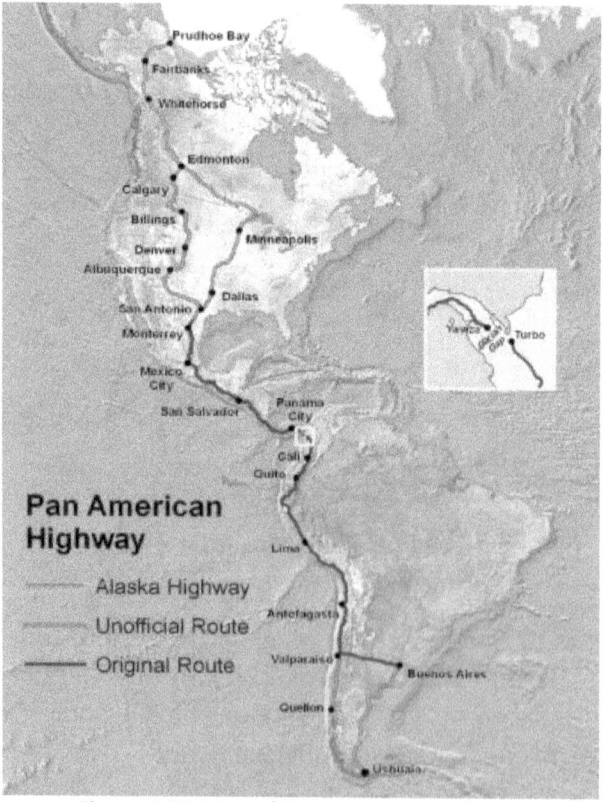

Figure 4.22 Map of Pan American Highway

That's 5 continents connected together. What about the 6th and 7th, Australia and Antarctica? Only a few scientists reside in

Antarctica, so there is no point to connecting them to the Global Maglev Network. It would be very desirable to connect Australia, but unfortunately, the ocean crossing distance is too great to be practical. Australia can still have its own National Maglev Network.

China and India will be of special importance in the Global Maglev Network. Their combined population is 30% of total World population. As described below, China's annual GDP (Gross Domestic Product) is projected to grow by a factor of 5 from today's (2014) value by 2050 AD, making it the #1 ranking in 2050 GDP. India's GDP is projected to grow by a factor of 10 by 2050 making it #3 in the ranking of 2050 national GDP's.

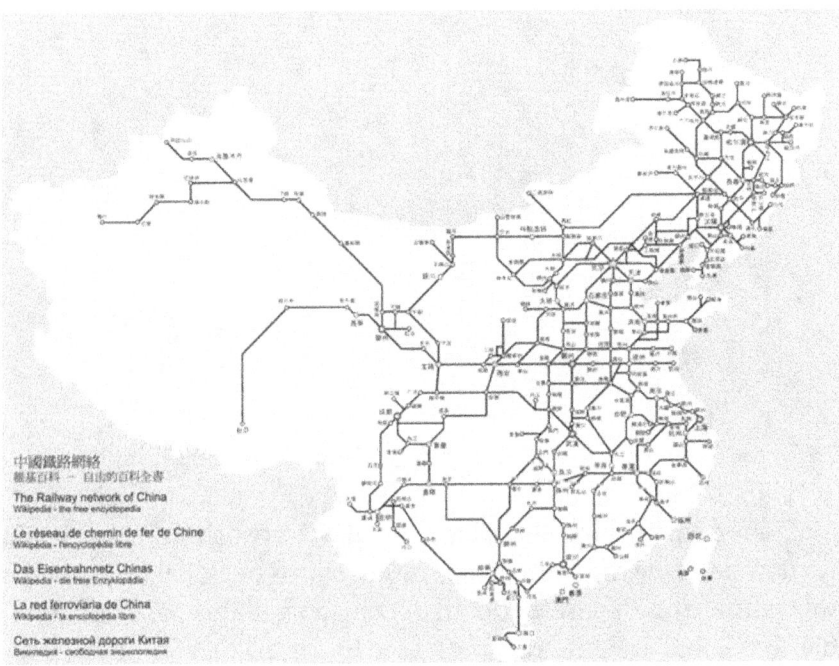

Figure 4.23 Map of Railway Network in China

China and India already have extensive national railway systems (Figures 4.23 and 4.24). Adapting them for Maglev travel, where practical, is a very attractive possibility, particularly in China, which has in the past few years, built many thousands of miles of

113

High Speed Rail (HSR) lines. Where appropriate, new Maglev guideway systems would be built.

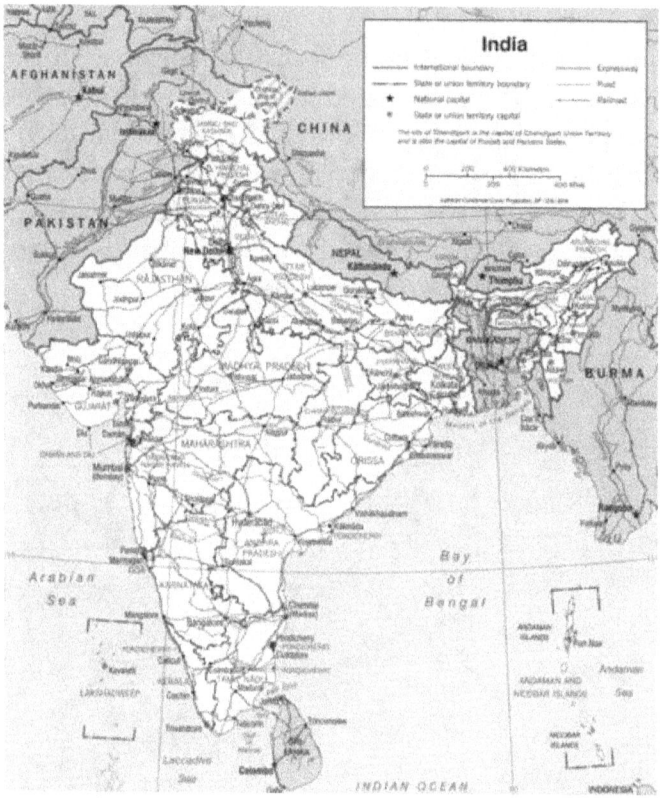
Figure 4.24 Map of India's Transport Networks

Before describing the Global Maglev Network in more detail, we now describe the present and future status of global transport, if we continue on the present path of transport that is based on fossil fuels. Global freight transport activity in 2050 will be 2 times greater than today (2014) and global passenger transport activity will be 2.6 times greater than today. (Table 4.9)

Along with the large increase in global transport activity will come a large increase in global carbon dioxide. Global $CO_2$ emissions from fossil fuels in 2014 were 32 Billion tonnes. Of that, approximately 27% came from transport, about 7.4 Billion tonnes.

For an R value of 2.1 for $CO_2$ emissions, 2050/2014, the global transport carbon dioxide emissions in 2050 would be 2.1 x 7.4 = 15.5 Billion tonnes, almost 1/2 of the present total (2014) emissions.

This is very important. One can shift from fossil fuels used for electrical generation to wind, ground solar, and especially beamed solar power as described in Chapter 5, and use the electrical energy from them for the industrial and residential sectors.

However, it is also very important to electrify transport if runaway global warming and ocean death are to be averted. As discussed light duty vehicles (LDVs)—that is, cars, SUVs, light trucks, together with highway freight trucks, buses and ships account for approximately 85% of transport $CO_2$ emissions, as shown in Table 4.10 for 2,000, 2030, and 2050. (3) Airplanes account for only about 12% of transport emissions.

Table 4.9
Index of Global Transport Activity, 2000 AD to 2050 AD
Reference: Transport Outlook 2011 (3)

| Year | Index | | |
|---|---|---|---|
| | Global Passenger Transport Activity (Passenger Kilometers) | Global Freight Transport Activity (tonne Km) | Global CO2 Emissions from Transport |
| 2000 | 100 | 100 | 100 |
| 2014 | 140 | 150 | 130 |
| 2020 | 170 | 170 | 150 |
| 2030 | 210 | 200 | 170 |
| 2040 | 270 | 250 | 230 |
| 2050 | 360 | 300 | 270 |
| Ratio 2050/2014 | 2.6/1 | 2/1 | 2.1/1 |

Global freight transport activity in 2050 will be 2 times greater than today (2014) and global passenger transport activity will be 2.6 times greater than today. (Table 4.9)

115

The biggest contributor to CO2 emissions from transport, approximately 1/2 of the total is Light Duty Vehicles – autos, SUVs and light trucks. This is not surprising. People love LDVs. They can go when they want, where they want, without other people crowded around them, at reasonable cost. If they want to travel long distances, hundreds of miles, they will consider taking airplanes or rail. If parking is not available or too expensive, they will take busses and other modes of public transit, but reluctantly.

The higher the per capita income in a country, the more motor vehicles per capita. With more money, people are more likely to buy one. Figure 4.25 illustrates the correlation between per capita GDP income of a country and motor vehicles per 1,000 passengers.

Table 4.10
Modal Composition of Global CO2 Emissions from Transport Vehicles Use (4)

| Mode | 2000 AD | 2030AD | 2050AD |
| --- | --- | --- | --- |
| Freight & Passenger rail | 2.3 | 1.9 | 1.5 |
| Buses | 6.3 | 4.3 | 3.0 |
| Air | 12.4 | 13.8 | 12.0 |
| Freight Trucks | 23.5 | 23.3 | 21.6 |
| LDVs | 42.5 | 45.2 | 52.1 |
| 2-3 Wheelers | 2.4 | 2.2 | 2.0 |
| Waterborne | 10.6 | 9.2 | 7.8 |
| | 100% | 100% | 100% |

A list of the 20 countries' GDP per capita (5) and motor vehicles per capita (6) is given in Table 4.11, together with their population and land area. It is clear that motor vehicles per capita only correlates with GDP per capita and not with population or land area. (7) Russia, with a land area of 17 million square kilometers, had 293 motor vehicles per capita compared to Mexico, which has 275 motor vehicles per capita, and an area of only 2 million square kilometers. Their GDP per capita's are close, $17,887 for Russia, and $15, 563 for Mexico. Similarly, India has a population of 1,170 million and 41 motor vehicles per capita while Vietnam has 23 motor vehicles per capita and a population of only 90 million.

Figure 4.25
Automobiles Per 1,000 Persons as a Function of GDP Per Capita for Different Countries, Basis: 2010 – 2013 Data PPP (Adjusted)

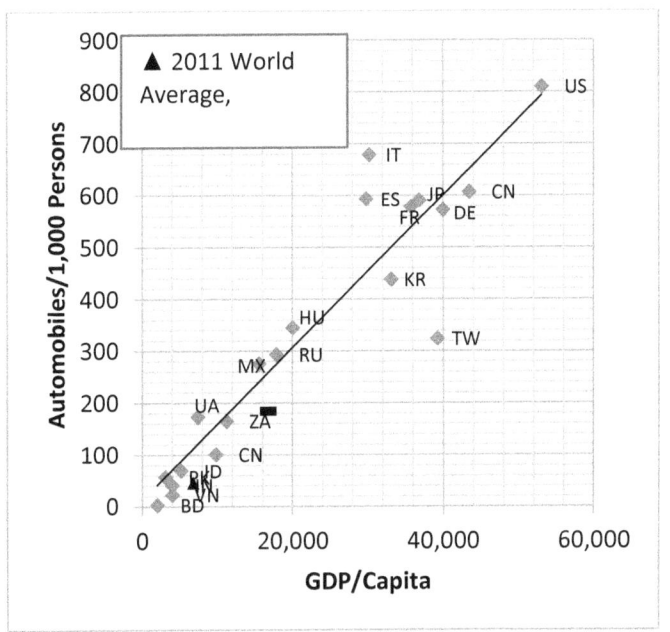

Of particular interest are how the World average motor vehicles per capita also correlate with World average GDP per capita. In 2011, World average GDP per capita was $10,000 per capita, with a World population of 7 Billion. In 2050 AD with 9 Billion people, World average GDP per capita is projected to be $22,000, a factor of 2.2 increase. With 1 Billion motor vehicles in 2014 and 2.5 Billion projected for 2050, the motor vehicle per capita will increase from 143 vehicles per 1,000 persons to 278 vehicles per 1,000 persons, slightly less than a factor of 2.

Table 4.11
GDP per Capita, automobiles per 1000 Persons, Population, and Area for
Different Countries, Basis 2010-2013 Data (PPP Adjusted)

| Rank | Country | Autos per Thousand Persons | GDP Per Capita in US$ | Population in Millions | Area in Millions of square KM |
|------|---------|---------|---------|---------|---------|
| 1 | US | 809 | 53,101 | 308 | 9.5 |
| 2 | Italy | 679 | 30,289 | 61 | 0.30 |
| 3 | Canada | 607 | 43,472 | 34 | 10.1 |
| 4 | Spain | 593 | 29,851 | 47 | 0.50 |
| 5 | Japan | 591 | 36,849 | 128 | 0.38 |
| 6 | France | 578 | 35,784 | 65 | 0.64 |
| 7 | Germany | 572 | 40,007 | 82 | 0.35 |
| 8 | S. Korea | 438 | 33,188 | 49 | 0.10 |
| 9 | Hungary | 345 | 20,065 | 10 | 0.09 |
| 10 | Taiwan | 324 | 39,267 | 23 | 0.04 |
| 11 | Russia | 293 | 17,887 | 139 | 17.1 |
| 12 | Mexico | 275 | 15,563 | 112 | 2.0 |
| 13 | Ukraine | 173 | 7,423 | 45 | 0.60 |
| 14 | S. Africa | 165 | 11,259 | 49 | 1.2 |
| 15 | China | 101 | 9,844 | 1,330 | 9.6 |
| 16 | Indonesia | 69 | 5,214 | 243 | 1.9 |
| 17 | Pakistan | 57 | 3,149 | 184 | 0.85 |
| 18 | India | 41 | 4,007 | 1,170 | 3.2 |
| 19 | Vietnam | 23 | 4,012 | 90 | 0.33 |
| 20 | Bangladesh | 3 | 2,080 | 156 | 0.15 |

We now examine how Maglev can meet the challenge of the quickly growing passenger and freight transport activities in countries with rapidly expanding economies. We also examine how Maglev can improve passenger and freight transport in already industrialized countries to be low in cost, more energy efficient, and more environmentally acceptable.

Table 4.12
Gross Domestic Product (GDP) of Words 10 Largest Economies in 2050 AD,
Compared to 2014 GDP Levels, Basis: Nominal GDP (Non PPP)

| 2050 Rank | Country | 2050 GDP in Trillion US$ | 2014 GDP in Trillion US$ | Ratio 2050GDP/ 2014GDP |
|---|---|---|---|---|
| 1 | China | 50.8 | 10.0 | 5.1 |
| 2 | United States | 38.0 | 17.5 | 2.2 |
| 3 | India | 21.0 | 2.0 | 10.5 |
| 4 | Japan | 7.8 | 4.8 | 1.6 |
| 5 | Brazil | 7.5 | 2.2 | 3.4 |
| 6 | Russia | 6.2 | 2.1 | 3.0 |
| 7 | Germany | 6.2 | 3.9 | 1.6 |
| 8 | United Kingdom | 5.8 | 2.8 | 2.1 |
| 9 | Mexico | 5.8 | 1.3 | 4.4 |
| 10 | France | 5.7 | 2.9 | 2.0 |

Table 4.12 lists the GDP's of the 10 Countries projected to have the largest economies in 2050 AD(8), and their present (2014) GDP's, together with the ratio of their 2050 GDP/2014 GDP.

- India has the largest growth ratio, R=21.0/2.0=10.5
- China is the next largest, with R=50.8/10.0=5.1
- The US, now the largest economy, is projected to slow, with R=38.0/17.5=2.2
- Japan has the lowest growth rate, with R=7.8/4.9=1.6
- Along with Germany, R=6.2/2.9=2.0, France is a bit better, with R=5.7/2.9=2.0
- Russian and Brazil are in-between, with R values on the order of 3, with Mexico at R=5.8/1.3=4.4, slight less than China

Their very large populations and large future economies make them especially significant in terms of future World transport. Later in the chapter, we examine in detail how Maglev will help meet their transport needs.

China and India are the biggest in growth and also the biggest in national populations (7).

|  | China | India | China + India | Total World | % of Total World |
|---|---|---|---|---|---|
| 2015 Populations in Billions | 1.36 | 1.25 | 2.71 | 7 | 39% |
| 2050 Population in Billions | 1.30 | 1.66 | 2.96 | 9 | 33% |

In summary, the projections for global transport and Global GDP in 2050 predict growth factors of 2 to 3 in travel activity, GDP per capita, numbers of motor vehicles, and carbon dioxide emissions from transport.

Such growth factors are not sustainable if transport, which is now primarily based on fossil fuels, continues to depend on the combustion of fossil fuels. Global warming, ocean acidification, and depletion of fossil fuel reserves will act, causing environmental and economic disaster.

For a sustainable, prosperous World with a better life for humanity, it is absolutely necessary to transform to electric transport and cease its dependence on fossil fuels, and to make the transition soon. Maglev will be a necessity in this inevitable global transition to electric transport.

The Global Maglev Network is described in the following 4 sections:

- Maglev Intercontinental Connections
- Maglev Systems for Countries in Each Continent
- Maglev Alternative to Ocean Shipping
- Global Maglev Schedule and Cost

*Turning to the first section, Maglev Intercontinental connections, there are 4 intercontinental Maglev connections.*

- Europe to Africa across the Strait of Gibraltar
- Asia to North America across the Bering Strait
- Europe to Asia through Siberia
- North America to South America through the Isthmus of Panama

## The Europe to Africa connection is across the Strait of Gibraltar

Figure 4.19 shows a satellite view of the Strait of Gibraltar. At its narrowest part, the strait is only 9 mile across. Spain and Morocco appointed a commission in 2003 to study the feasibility of a tunnel under the Strait of Gibraltar's sea floor. The proposed tunnel would be 25 miles long and 980 feet deep. It would be very expensive. The English Chunnel cost approximately 10 Billion dollars, when it was built in 1985, and would probably be twice that if built today. Moreover, it would be much deeper than the Chunnel, with less favorable geology and earthquakes.

The Maglev Europe to Africa crossing could be made in 2 one-way underwater tunnels like the Chunnel or more likely, in underwater floating tubes of reinforced polymer concrete, which is much stronger than ordinary concrete. The tubes would have positive buoyancy, and be anchored to the sea floor beneath by cables. They would be located at a depth of approximately 100 meters (300 feet) to minimize wave action.

The Maglev crossing would be approximately 14 miles long, running from Tanya in Spain to Tangier in Morocco. The location chosen reduces the length of the anchor cables from the maximum water depth of 1,000 meters to the Seafloor for the 9 mile wide crossing to about 500 meters. Capital cost would be in the range of 1 to 2 Billion dollars.

With consists of 3 Maglev vehicles coupled together, each carrying 100 passengers with 1 minute headings between consists, the Maglev Gibraltar crossing could transport 320 million passengers per year, 20 times the 16 million passengers traveling through the English Chunnel. Clearly, this passenger capacity far exceeds the likely traffic, but it does allow for large growth in the Gibraltar crossing passenger traffic.

Similarly, with a freight loading of 40 tons per Maglev vehicle, 3 vehicles per consist, and 1 minute heading between consists, the 2 freight tubes in the Gibraltar crossing could transport 120 million tons of freight per year, 12 times greater than the 10 million tons transported by the English Chunnel.

Only a small number of Maglev vehicles will be required for the Gibraltar crossing. At 200 mph, the Maglev vehicles consist would take only 4 minutes to travel the 14 mile tube system between the 2 terminals in Spain and Morocco. Allowing for peak demand periods and maintenance, a fleet of 30 Maglev passenger vehicles and 20 Maglev vehicles could easily handle traffic flow volumes well in excess of the passenger and freight volumes currently transported through the English Chunnel. The cost of the 50 Maglev vehicle fleet would be well under 1 Billion dollars.

The Asia to North America Connection across the Bering Strait.(Figure 4.21) This crossing is longer, 50 miles in length, compared to 14 miles for the Gibraltar crossing, but the maximum water depth is much less, 180 feet versus 3,000 feet for the Gibraltar crossing.

First proposed in 1890 as a Bridge across the Strait by William Gilpin, the first Governor of Colorado Territory, it has been advocated by many since then. Joseph Straus, the future designer of the Golden Gate Bridge, proposed a Bering Strait Bridge in 1892 in his senior thesis. In 1905 Tsar Nicholas II approved a plan for a tunnel across the Strait. World War I and Lenin stopped the project. The latest proposal is for a 64 mile road and high speed rail tunnel. Russian's Prime Minister Putin has approved a plan to build a railroad to the Bering Strait that would connect to the proposed TKM-World Link crossing. The projected cost for the Bering Strait tunnel crossing is 65 Billion dollars, over 6 times the cost of the 31 mile English Chunnel, which cost about 10 Billion dollars in 1985 dollars. However, Russia did not promise to supply the funds.

The Bering Strait crossing would use the same approach as the Strait of Gibraltar crossing, prefabricated reinforced polymer concrete tubes anchored to the seabed beneath. The principle difference between the Bering Strait and the Strait of Gibraltar crossing would be the longer length of the Bering Strait crossing, 50 mile miles compared to 14 miles for the Gibraltar crossing, and a shallower ocean – maximum depth of 180 feet, compared to 1500 feet.

The shallow depth enables the floating Maglev tunnel to be anchored only a few feet about the sea bed, reducing the cost of anchoring, and making the assembly process easier and cheaper. The total projected construction cost for the 50 mile long 4 tube system – 2 tubes going east across the Strait and 2 tubes going west – would be 6.4 Billion dollars. Amortizing the 6.4 Billion dollars over 30 years, with an annual flow of 10 million TEU freight containers across the Bering Strait connection, corresponds to only 20 dollars per TEU, compared to the 3,000 dollars cost to transport a TEU container by ship.

The Bering Strait crossing would connect the Russian side of the Strait to the Europe area Trans-Siberia Maglev Line described below. On the Alaska side of the crossing it would connect to the Pan American Maglev Highway, which would run from Alaska's Seward Peninsula down through Canada to the US National Maglev Network described earlier, and from there through Mexico and Central America, connecting to the South American continent through the Isthmus of Panama. From there, the Pan American Highway runs down the West Coast of South America to its southern tip.

## Europe to Asia connection through Siberia

Figure 4.20 shows a map of the present Trans-Siberian railway that runs from Moscow to Vladivostok. The Trans-Manchurian line branches off from the Trans-Siberian Rail at Taranga, 3,898 miles from Moscow, and continues on for another 1,670 miles to Beijing

123

in China. The Trans-Mongolian line branches off from the Trans-Siberian Railway at Ussuriysk, 5,680 miles from Moscow, and continues on for 700 more miles to Pyongyang in North Korea. (9)

Passenger trip time from Moscow to Vladivostok (5,772 miles) is 6 days and 4 hours at an average speed of 39 mph. The fastest freight trip time with good coordination from Beijing to Hamburg, Germany or Europe is about 16 days, for an average speed of approximately 15 mph. Average freight trip-time is about 25 days for an average speed of approximately 10 mph.

Russia plans investing 11 Billion dollars to reduce the cargo trip time across Russia from its Pacific ports to its border with Europe down to 7 days. However the Trans-Siberian Railway and other Russian rail lines have a railway gauge problem. Russia's railway gauge is broader than the standard 4 foot 8 inch rail gauge used in China and Europe. As a result, cargo traveling from Europe to Asia and vice versa by rail through Russia has to be unloaded & loaded 2 times – at Russia's Western and Eastern borders. This slows down the average transport speed and increases trip time.

A high speed Maglev route for cargo and passengers across Russia would in general follow the Trans-Siberian railway route. Pre-fabricated Maglev guideway beams and piers would be transported on the existing railway line to assembly sites, there to be erected by cranes on already prepared footings, as planned for the U.S. National Maglev Network.

China to Europe in 1 Day -- not 7 days by rail, or 35 days by ship. And cheaper than by rail or ship. And environmentally benign – no greenhouse gas emissions, no pollutants from dirty bunker fuel emissions from ships that kill many thousands of people annually. Companies that make goods and the companies that buy the goods will love the Trans-Siberian Maglev Line.

The National Maglev Networks in the various countries in Europe and Asia will be connected to the Trans-Siberian Maglev Line, enabling goods and passengers from one country in Europe to go to their destination in an Asian country. For example, from Germany, travelers and cargo would use a Maglev line from Germany to the Maglev line through Poland to connect to the Trans-

Siberian Maglev line through Serbia, then leave Russia at the Russia-China Border, traveling from there to their destination in China on a Chinese Maglev line. Potential Maglev routes inside the individual countries that connect to the Trans-Siberian Maglev line are described later.

The Trans-Siberian Maglev line will have a total 2-way route mileage of approximately 10,000 miles, taking into account the approximately 5,800 main route mileage between Moscow and Vladivostok plus another 4,200 miles for the various Maglev branch routes that connect to the main route. The total construction cost for the 10,000 mile system would be about 300 Billion dollars. World annual GDP is projected to be more than 190 Trillion dollars in 2050 AD. Over the next 30 years at an assumed average of about 100 Trillion dollars yearly, World GDP would total approximately 3,000 Trillion dollars, 10,000 times greater than the 300 Billion dollars construction cost. Moreover, the savings in transport costs will far exceed its construction cost.

We turn now to the 4th intercontinental Maglev connection:

## The North America to South America connection across the Isthmus of Panama

Shown in Figure 4.22, the Pan American Maglev Highway would run from the Seward Peninsula in Alaska down to the tip of South America. Total mileage for the Pan American Maglev Highway, including both the official and unofficial routes shown in Figure 4.22 is approximately 30,000 miles, a bit longer than the US National Maglev Network described previously. Deducting the approximately 2,000 miles in the US National Maglev Network that would be part of the Pan American Maglev Highway, its total mileage would be about 28,000 miles.

The Pan American Highway directly connects almost all of the countries in North and South America. In North America, it serves:

| | |
|---|---|
| Canada | Honduras |
| The United States | Nicaragua |
| Mexico | Costa Rica |
| Guatemala | Panama |
| El Salvador | |

The only North American countries not directly connected to the Pan American Maglev Highway are Belize and British Honduras. In South America the following countries are directly served by the Pan American Maglev Highway.

| | |
|---|---|
| Suriname | Ecuador |
| Guyana | Peru |
| Brazil | Chile |
| Venezuela | Argentina |
| Columbia | |

Maglev branches would connect Paraguay and Uruguay to the Pan American Maglev Highway.

## North America, South America, Asia, Europe, and Africa

We now take a look at how the various countries inside the 5 continents are connected to the Global Maglev System. Rather than each continent separately, we look at 3 continental systems:

The Pan American Maglev System, connecting the countries in North and South America, from Alaska to Tierra del Fuego, would connect to Asia via the Bering Strait crossing.

The Pan African Maglev System would encircle Africa, directly connecting all of its countries on the East, West and North Coasts of the continent. Island countries would be connected to the Pan African Coastal systems by branch Maglev Systems. Connection to the Pan Europe – Asia Maglev system would be made at 2 points – by the Gibraltar Strait Crossing on Africa's West Coast, and by the Isthmus of Suez crossing on Africa's East Coast.

The Pan Europe – Asia Maglev System would travel from Britain in the West of the Europe-Asia continent—it really is one continent, with the division between Europe and Asia being cultural, not geographical – to China on the East of the continent. The Pan Europe – Asia Maglev System connects to the Pan American Maglev System at 2 points – the Gibraltar Strait crossing in the West, and the Isthmus of Suez crossing in the East.

To which "continent" do certain countries belong? There is general agreement that France is part of "Europe", for example, and China is part of "Asia". What about Russia? Some put it in Europe, some in Asia. Most definitions split it two –"European Russia", and "Russian Asia". The split seems rather arbitrary and is defined differently by different experts. Most use the Ural Mountains and other physical features to define European and Asian Russia. Generally about 80% of Russia's population lives in "Europe". And Turkey? Some put it in Europe and some in Asia. It makes no real difference. The Pan Europe – Asia Maglev System considers them as an integrated entity, with virtually all of the countries in "Europe-Asia" corrected by Maglev.

## The Pan-American Maglev System

The Pan American Maglev Highway, (Figure 4.22) follows the route proposed for the Pan American Highway. (10) With side branches, it connects countries in North and South America.

North America, principally due to the US economy, is much wealthier than South America, with total nominal GDP of 20.2 Trillion dollars annually, compared to only 4.1 Trillion dollars for South America, a factor of 5 difference. North America has a somewhat greater population, 514 million than South America's 404 million, but the population difference is much smaller than the GDP difference.

South America's continental economy is probably too small to fully finance its portion of the Pan American Maglev System. Some investment from wealthier North America will probably be necessary.

Total combined annual 2013 GDP is approximately 25 Trillion dollars. By 2050 AD, total annual GDP will be approximately 50 Trillion dollars, measured in today's dollars. The accumulated continental GDP from today to 2050 AD, 35 years from now, will be more than 1,000 Trillion dollars, a thousand times greater than the approximately Trillion dollars investment to create the Pan American Maglev System.

So, on a continental basis, there clearly are ample funds to build the Pan American Maglev System. The economic benefits from it will far exceed the required investment, both in savings from lower cost of transport, and also from the opportunities it will enable the developed countries in North American and South America to grow their economies. Table 4.13 summarizes the parameters for the Pan American Maglev System.

Table 4.13 Summary of Pan American Maglev System

- Northern Terminus: Bering Strait, Alaska
- Southern Terminus: Tierra del Fuego, Argentina
- Countries Directly Served in North America: US, Canada, Mexico, Guatemala, Belize, Honduras, El Salvador, Nicaragua, Costa Rico, Panama
- Countries Directly Served in South America: Columbia, Venezuela, Ecuador, Peru, Bolivia, Chile, Argentina, Uruguay, Brazil, Paraguay, Guyana, Suriname
- North America Population Directly Served: 514 Million
- South America Population Directly Served: 404 Million
- 33,000 Route Miles of 2-Way 300 mph elevated Maglev Guideway
- Maglev System transports Passengers, Freight, Trucks and Autos
- Investment for 33,000 miles of Maglev Routes: 1 Trillion Dollars
- Present Annual GDP of North & South America: 25 Trillion Dollars
- Projected Cumulative GDP, 2015 to 2050: $1,300 Trillion Dollars

The Pan American Maglev System routes shown in Figure 4.22 use the 2nd Generation Maglev-2000 elevated monorail guideway that carries passengers, trucks and autos at 300 mph. A more detailed description of the 2nd Generation System and its capabilities is given earlier on the US National Maglev Network. The projected route mileage of the Pan American Highway route is 30,000 miles. This includes approximately 2,000 miles of the US continental National Maglev Network. Deducting this portion results in approximately 28,000 miles for the Pan American Maglev system. However, approximately 5,000 miles of 300 mph Maglev guideway is added to the total mileage, for the extension from Fairbanks to the Bering Strait crossing route, plus a route across Canada to join its West Coast cities to its East Coast cities. Total mileage for the Pan American Maglev System is then approximately 33,000 miles.

The projected construction investment for the 300 mph 2nd Generation Maglev-2000 elevated guideway in the US and Canada is 30 million dollars per 2-way mile. The cost may be less in the countries with lower labor costs and smaller GDPs that connect to the Pan American Maglev System. However, for this study, to be conservative, we take the investment to be constant at 30 million dollars per 2-way mile, for all the countries in the Pan American and Maglev Systems.

At 30 million dollars per 2-way mile, total investment for the 33,000 mile Pan American Maglev system would be about 1 Trillion dollars. Not included are the investments for stations and Maglev vehicles which will be about 100 Billion dollars, making the total investment on the order of 1.1 Trillion dollars.

Also not included are the internal investments that would be undertaken by the various countries to provide good access to the Pan American Maglev System that runs through their country. In general, the 300 mph Pan American System would service the principal cities in each country. A country would upgrade its roads and railways to provide quicker and easy access to the Pan American Maglev System at those cities.

In particular, countries with extensive conventional steel wheel on rail systems could upgrade their railways, by adapting the trackage for Maglev travel at a low investment of approximately 4 million dollars per one-way mile, as described in earlier for US commuter rail and subways.

For example, Mexico, Brazil and Argentina have extensive rail networks but move relatively small numbers of passenger miles and freight ton miles per year. (10)

| Parameter | Mexico | Brazil | Argentina |
|---|---|---|---|
| Rail Mileage (miles) | 26,704 | 28,538 | 39,966 |
| Passenger Miles Per Year (Billions) | <3 | <3 | 5 |
| Freight Ton Miles Per Year(Billions) | 41 | 160 | 7 |

In comparison, the US transports 1,500 Billion tons of rail freight annually, 10 times more than Brazil, Mexico and Argentina. This is not surprising given America's greater population and higher GDP. However, as the population and GDP of the other countries connected to the Pan American Maglev System increase in the years ahead, having a substantial internal railway system already in place that can be adapted at low investment to Maglev, enabling fast and easy access to the Pan American Maglev System will be very advantageous.

Overall conclusions. The Pan American Maglev System will directly serve all of the countries in North America and South America, providing fast, efficient, low-cost travel for passengers, freight, and motor vehicles. Projected investment for the 33,000 mile Maglev System, which will run from the Bering Strait Maglev crossing that connects North America to Asia, down to the Tierra del Fuego at the bottom of South America, is approximately 1 Trillion dollars.

The present total annual GDP for North and South America is approximately 25 Billion dollars, and is expected to double by 2050 AD, 35 years from now. The total accumulated GDP for North and South America from now to 2050 AD will be on the order of 1300 Trillion dollars, more than 1,000 times greater than the investment for the Pan American Maglev System. The economic earnings and growth enabled by the Pan American Maglev system will be far greater than the investment for it.

## The Pan-African Maglev System

Africa's present population (2013) of 1.1 Billion people is projected by the UN to grow to 2.4 Billion in 2050, making it 29% of the World's 9 Billion in 2050. (7) But Africa is a very poor continent, with a present annual GDP of only 2.6 Trillion dollars – 3.6 percent of the World total annual GDP of 72 Trillion dollars.(8) The average per capita GDP for Africans is only 1600 dollars, compared to the World average of about $10,000 per capita – about 1/6th of the World average.

The total World economy is projected to grow by about a factor of 2.6 by 2050, with a total annual GDP on the order of 200 Trillion dollars. The average GDP per capita in 2050, based on 9 Billion people and a 200 Billion dollar annual GDP would be about $22,000 per person,(in constant dollars), more than 2 times the present World average GDP per capita.

Even to maintain its present value of 1/6th of World GDP per capita, Africa would need a per capita value of 1/6 x $22,000 = $3,700 per person. With a growth to 2.4 Billion population, that means Africa would have to have a GDP of 2.4 Billion x $3,700 = 8.9 $Trillion in 2050 – almost 4 times its present GDP.

Whether or not such a GDP growth is possible for Africa is unknown. What is known is that without a modern transport system, it will not be possible. Most of Africa's 58 countries have less than 1,000 miles of railway tracks. Only 6 countries have more than 2,000 miles. South Africa has the most mileage (12,000 miles), followed by Egypt (4,000 miles), Sudan (3,300 miles), Algeria (2,500 miles), Democratic Republic of Congo (2,400 miles) and Nigeria (2,000 miles).(10)

African transport is constrained not only by its very limited railway and highway mileage but in the railway sector by the wide variety of different railway gauges in Africa's countries. In South

Africa, almost all railways use the Cape narrow gauge (3 feet 6 inches). Only a small fraction of its railway are standard gauge (4ft 8.5 in), the widely used standard for most of the World. 35% of South Africa's 12,000 rail mileage carries no activity or very low activity. (11)

Many of Africa's countries operate with Cape Gauge (3 feet 6 inches), including Angola, Botswana, Congo, Ghana, Mozambique, Namibia, Nigeria, Sudan, Zambia, and Zimbabwe. Other countries use the narrower metric gauge (1,000 mm, or 3 feet 3 3/8 inches), including Kenya, Uganda, Ethiopia, Cameroon, and Tunisia. There are even narrower gauges, down to as little as 2 feet. Basically, each country operates transport individually, with very little interconnection. (12)

Table 4.14
List of African Countries and Populations Connected to the Pan-African Maglev System, Northern Coast Route – Morocco to Egypt

| 5 Countries Directly Connected (Population in Millions) | 3 Countries Indirectly Connected (Population in Millions) |
|---|---|
| Morocco (32.9), Libya (6.3) | Mali (16.7) |
| Algeria (38.3), Egypt (34.6) | Niger (17.5) |
| Tunisia (10.9) | Chad (12.9) |
| Total Population 173 Million | Total Population 47 Million |

Maglev can connect the countries of Africa into a continental network for high speed, efficient, low cost transport of passengers, freight, trucks, and autos. Figure 4.26 shows a map of the African continent encircled by the Pan-African Maglev system that runs along its North, East, and West Coasts. Table 4.14 lists the countries and their populations that are directly connected to the North, East and West Coast routes that run through their countries. Of Africa's 58 countries 36 are directly connected.

(Table 4.14 continued)

| 12 Countries Directly Connected (Population in Millions) | 16 Countries Individually Connected (Population in Millions) |
|---|---|
| Sudan (35.2), Tanzania (45.9) | S. Sudan (10.3), Cen. Afr Republic (5.2) |
| Eritrea (5.0), Mozambique (24.5) | Dem.Rep.Congo (74.6), Uganda (35.4) |
| Djibouti (0.4), Swaziland (1.1) | Burundi (9.0), Zambia (14.1) |
| Ethiopia (86.6), Lesotho (1.9) | Zimbabwe (13.1), Rwanda (10.8) |
| Somalia (9.6), Kenya (43.3), | Malawi (15.3), Botswana (2.1) |
| South Africa (53.0), Cape Verde (0.31) | Madagascar (21.8), Seychelles (0.09) |
| | Mayotte (0.22), Comoros (0.74) |
| | Mauritania (1.3), Reunion (0.86) |
| **Total Population 307 Million** | Total Population 215 Million |

Another 14 countries on the African continent (Table 4.14) are indirectly connected to neighboring countries through which the Pan-African Maglev System runs by railways and/or highways. Maglev branches can be built to connect these countries to the main Pan-African Maglev System. An alternative possibility is to adapt existing railway lines for Maglev operation by placing panels of aluminum loops on the crossties of the railroad trackage, as described earlier. Adaptation to Maglev can be done on any of the various railway gauges, since the rails do not contact the traveling Maglev vehicles. The operating Maglev vehicles on the adapted railway tracks can electronically switch onto and off from the Mainline Pan African Maglev System and travel to whatever destination in Africa they want to. The investment for adapting existing railway tracks for Maglev operation is low, or the order of 4 million dollars per one-way mile.

(Table 4.14 continued)

| 19 Countries Directly Connected (Populations in Millions) | 3 Countries Indirectly Connected (Populations in Millions) |
|---|---|
| Namibia (2.2), Guinea Conakry (11.8) | Burkina Faso (17.3) |
| Republic of Congo (4.5), Gabon (2.2) | Sao Tome (0.19) |
| Equatorial Guinea (1.8), Benin (9.7) | Saint Helena (0.004) |
| Cameroon (20.9), Togo (6.7) | |
| Nigeria (177.1), Ghana (26.4) | |
| Liberia (3.9), Western Sahara (0.65) | |
| Mauritania (3.5), Senegal (13.6) | |
| Gambia (1.8), Guinea-Bissau (1.7) | |
| Ivory Coast(23.9), Angola(21.2) | |
| Sierra Leone (5.8), | |
| Total Population 389 Million | Total Population 17 Million |

8 of the 58 African countries, e.g., Madagascar, Seychelles, etc. are in the Indian and Atlantic ocean, too distant from the African continent for a bridge or tunnel connection. They would have to connect by ship. In general, their total population is small, about 25 million, a little over 2% of Africa's total population of 1.1 Billion. Overall, 75% of Africa's population is directly served by the Pan-African Maglev System, with 25% indirectly served. Table 4.15 summarizes the parameters for the Pan-African Maglev Systems. Total route mileage is approximately 18,000 miles, with 3,000 miles for the North Coast Route, 7,000 miles for the East Coast Route, and 8,000 miles for the West Coast Route. Total investment for the 18,000 mile system is 0.54 Trillion dollars, approximately 1/400th of the projected accumulated African GDP from 2015 to 2050.

Table 4.15 Summary of the Pan African Maglev System

- North Coast Maglev Route – Morocco to Egypt
- Strait of Gibraltar West Terminal, Isthmus of Suez East Terminal 3,000 Route Miles
- 5 Countries Directly Connected, 173 million population
- 3 Countries Indirectly Connected, 47 million population
- East Coast Maglev Route – Egypt to South Africa
- Isthmus of Suez North Terminal; Cape Horn South Terminal 7,000 Route Miles
- 12 Countries Directly Connected, 307 million population
- 19 Countries Indirectly Connected, 215 million population
- West Coast Maglev Route – South Africa to Morocco
- Cape Verde South Terminal; North Terminal, Strait of Gibraltar 8,000 Route Miles
- 19 Countries Directly Connected, 389 million population
- 3 Countries Indirectly Connected, 17 million population
- 75% of Africa's 1.1 Billion population directly connected to Pan-African Maglev System 18,000 Route Miles of 2-Way, 300 mph Elevated Maglev Guideway
- Maglev System Transports Passengers, Freight, Trucks, and Autos.
- Investment for 18,000 miles of Maglev Routes: 0.54 Trillion Dollars
- Present Annual GDP of Africa: 2.6 Trillion Dollars
- Projected Cumulative GDP 2015 to 2050: 230 Trillion Dollars

As with the Pan-American Maglev system, the transport savings enabled by the proposed Pan-African Maglev System will be much greater than the investment to build it.

Overall conclusions? The Pan-African Maglev System will directly serve 75% of Africa's population, who live in the 36 countries that the Maglev System passes through. The neighboring countries can connect to the Maglev System using existing railways and highways. With a low investment, existing railway trackage can be adapted for Maglev transport by attaching simple panels of

135

aluminum loops to the railway crossties, enabling Maglev vehicles to switch on to or off from the mainline system to the adapted Maglev trackage.

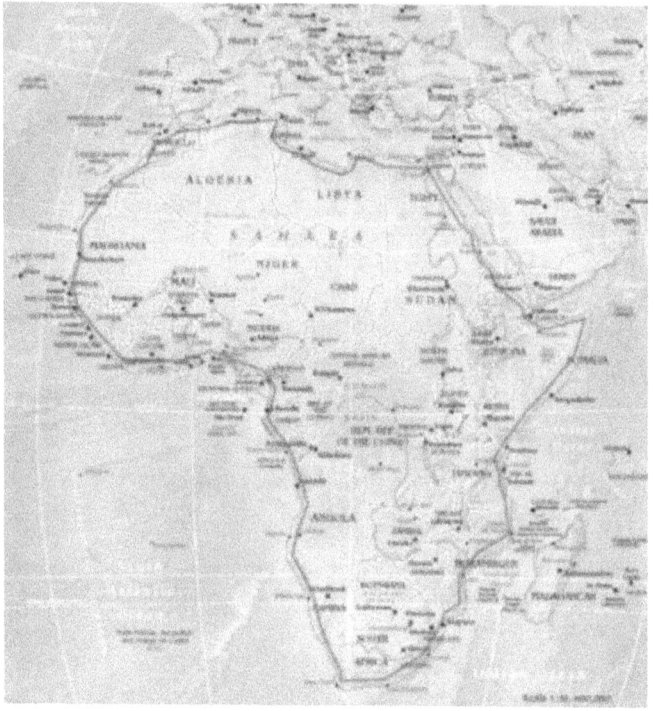

Figure 4.26

The investment for the Pan-African Maglev System is approximately ½ Trillion dollars, a small fraction of about 1/400th of the projected cumulative African GDP for the years 2015 to 2050. The economic savings by the Pan-African Maglev System will be much greater than the investment for it.

## The Pan Europe-Asia Maglev System

Europe and Asia are not physically separate continents, but rather one giant continent divided by culture, not geography. When historians, economists, marketers, etc. try to divide the one giant continent into 2 separate entities, there's a lot of disagreement.

Where is the boundary? Is Turkey in Asia or Europe? Where do you put Russia? Are the Ural Mountains the boundary? And so on.

Since the aim is to connect all the World's continents into a global transport network, the distinction between Europe and Asia is irrelevant. The goal is to have the Western terminus of Pan Europe-Asia Maglev System in Britain go through the Chunnel to France, cross "Europe" to Russia and virtually all of the "Asian" countries to the Pacific Ocean and the Bering Strait, where the Pan Europe-Asia Maglev System would connect with the Pan American Maglev System.

And not to leave Africa out, the Pan Europe-Asia Maglev System would connect with the Pan African Maglev System across the Strait of Gibraltar and at the Isthmus of Suez.

Europe and Asia have connected together for thousands of years for travel of goods and people. In ancient times, it was the Silk Road. Then came ships, Vasco de Gama and others.

Then came modern air travel. Now most freight travels by ship and most passengers by air. Very little travel by rail: who wants to spend 6 days and 4 hours traveling on the Trans-Siberian Railway from Moscow to Vladivostok? The Silk Road is now a World Heritage Site, honored in memory, but not in practice.

But that will change in the years ahead when the Pan Europe – Asia Maglev System the "New Silk Road" begins operation. Asia GDP is rapidly growing. Today, China's GDP is 10 Trillion dollars annually. In 2050, just 35 years from now, it is projected to be 50 Trillion dollars, measured in constant 2014 dollars. (8) India's GDP is 2 Trillion dollars annually, today, and projected to grow to 21 Trillion dollars annually in 2050. The GDP of other Asian countries will also grow by large factors. Europe GDP will grow but its growth factor will be much less.

Together, the combined GDP of Europe and Asia will be more than 70% of total World GDP in 2050 (8), with enormous amounts of goods and people moving back and forth between the countries of the "Europe-Asia" continent.

Where will the "New Silk Road" go? The Trans-Asian Railway (TAR) Project carried out by the United Nations Economic and Social Commission for Asia and the Pacific (UNESCAP) has studied this question in detail. Figure 4.27A shows a map of the TAR Project. (13)

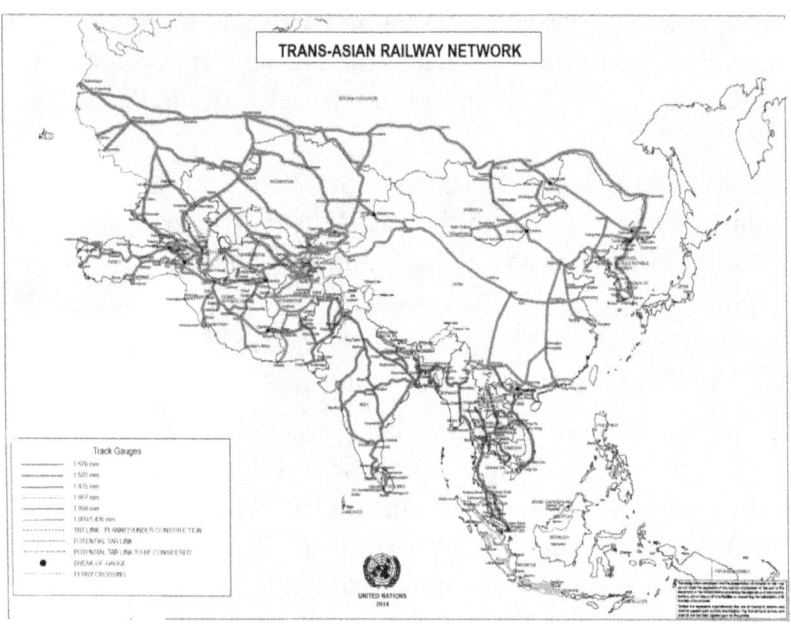

Figure 4.27A Map of Existing Trans-Asian Railway Network

The TAR Project is based on joining existing railway lines to form a network that would serve 26 countries in Asia. Some short railway links would have to be constructed, but almost all of its 81,000 Kilometer (50,000 mile) railway length is already there.

The Trans-Asian Railway Network Agreement was signed by 17 Asian nations on November 10, 2006 and formally came into force on June 11, 2009. Some progress has been made to implement the TAR Project, but the wide range of railway gauges used by the various countries in the TAR Project complicate things.

As in Africa, there are many different gauges in the TAR network, China, Iran, and Turkey use standard 4 feet 8.5 inch

138

gauge, Russia uses 4 feet 11 and 27/32 inch gauge, India and Pakistan use 5 feet 6 inch gauge, Bangladesh and Vietnam use 3feet 3/8 inch gauge, and Indonesia and Japan use 3 feet 6 inch gauge. Joining all the different railway gauges into a seamless system will be very difficult.

The Pan Europe-Asia Maglev System, besides providing 300 mph service for passengers, freight, trucks and autos instead of 30 mph railroad trains for passenger and freight, also solves the railway gauge problem, by adapting at low investment, all the different gauge systems to universally use Maglev.

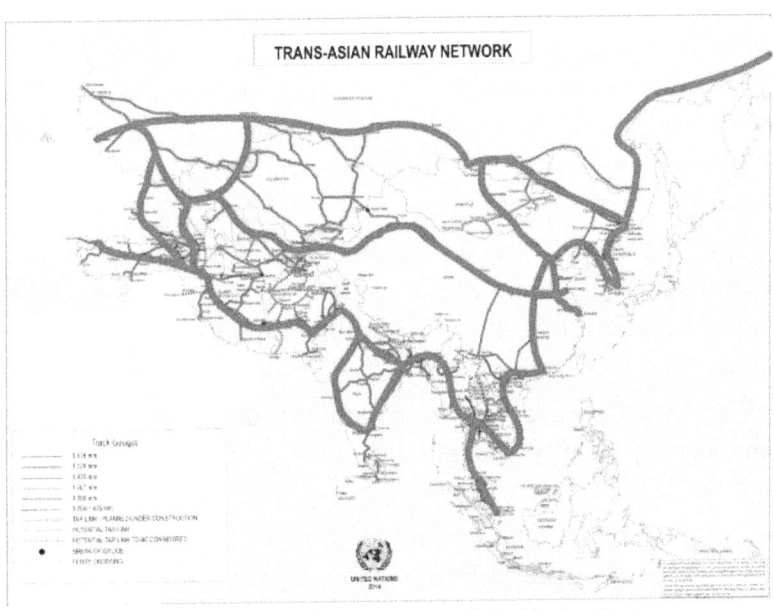

Figure 4.27B Map of Pan Europe – Asia Maglev System

Figure 4.27B shows the layout of 3 transcontinental routes of 300 mph elevated monorail guideway across Asia, that parallel the 4 principal corridors of the TAR Network and the countries they serve.

- North-East Asia:  China, Democratic People's Republic of Korea, Mongolia, Republic of Korea, Russian Federation. This route would parallel the Trans-Siberian Railway (Figure 4.20).

- Central Asia and Caucasus: Armenia, Azerbaijan, Georgia, Kazakhstan, Kyrgyzstan, Tajikistan, Turkmenistan, Uzbekistan.
- South-East Asia: Cambodia, Indonesia, Malaysia, Miramar, Singapore, Thailand, Vietnam.
- South Asia and Islamic Republic of Iran and Turkey: Bangladesh, India, Islamic Republic of Iran, Pakistan, Sri Lanka, Turkey.

The total route length of the TAR Network is 50,000 miles. The total route length of the 3 transcontinental, 300 mph elevated monorail routes is approximately 30,000 miles. The remaining 20,000 railway miles of the TAR Network would connect to the 300 mph Maglev routes using the existing conventional railways. Certain sections of the 20,000 miles of existing railway trackage could be adapted at very low investment to use Maglev, though at lower speeds than 300 mph.

The Pan Europe – ASIA Maglev System will have additional 300 mph elevated Monorail routes to the 30,000 miles in the TAR Network. These additional 300 mph routes would be built in 3 regions.

The Russian-Far East, from Vladivostok on the Pacific Ocean to the Bering Strait crossing where it would connect to the Pan American Maglev System.

Europe, where there would be 3 elevated 300 mph Maglev routes: across Northern Europe, Central Europe, and Southern Europe, connecting to the North-East Asia Maglev Route through Russia, and the South Asia, Iran, Turkey Maglev Route (Figure 4.27B). Virtually all of Europe's 50 countries would connect to the Pan Europe-Asia Maglev System.

The Middle East, where 300 mph Maglev routes would serve Iraq, Saudi Arabia, Yemen, Syria, United Arab Emirates, Israel, Jordan, Palestine, Lebanon, Oman, Kuwait, Qatar, and Bahrain. [Egypt is served by the Pan-African Maglev System, while Turkey and Iran are served by the Pan Europe – Asia Maglev System]. The

Middle East Maglev route would connect with the Pan-African Maglev System at the Isthmus of Suez.

The additional 300 mph elevated Maglev guideway route mileage in the 3 regions described above is approximately 10,000 miles. The total mileage for the Pan Europe – Asia Maglev System to 30,000 miles (TAR routes) plus 10,000 miles, equal to 40,000 miles At 30 million dollars per 2-way mile for the Maglev Guideway, this is a total of 1.2 Trillion dollars.

To put the investment for the Pan Europe – Asia Maglev System in perspective, consider the population and GDP of the countries it will serve, as shown below:

Cumulative GDP for Europe and Asia from 2015 to 2050 is projected to be more than 2100 Trillion dollars, 1700 times greater than the Pan Europe-Asia Maglev System investment

4.9 billion people would be served by the Pan Europe-Asia Maglev System. 70% of Word's present population of 7 Billion.

| Populations, Billions | | Annual GDP (Trillion USD) (constant $) | |
| --- | --- | --- | --- |
| Continent | 2014 | 2014 | 2050 (projected) |
| Europe | 0.74 | 18.5 | >31 |
| Asia | 4.2 | 24.4 | >94 |
| Total | 4.94 | 43 | >125 |
| Total World Pop. | 7.0 | Total World 73 | >188 |

The annual GDPs for 2050 are taken from the IMF projections of the 20 top economies for that year.(8) Total World GDP will be greater. The projections show that the GDP of Europe and Asia will be an even larger fraction of World GDP in 2050 on the order of 66%, than the already large fraction, 60%, that they are today.

Based on an average annual GDP of 60 Trillion dollars for Europe plus Asia over the 35-year period between 2015 and 2050 which is probably conservative, the accumulated GDP would be 2,100 Trillion dollars, 1,700 times greater than the 1.2 Trillion dollar investment for the 40,000 mile Pan Europe – Asia Maglev System.

Table 4.16 summarizes the principal parameters of the proposed Pan Europe – Asia Maglev system. It will enable a 21$^{st}$ Century Silk Road that will greatly benefit 70% of the World's population, increasing their living standards. The investment required is a very tiny fraction of the GDP of the countries that Pan Europe – Asia Maglev System will serve. It can be built in a relatively short time, following in many regions, the routes of the historic Silk Road, transporting millions of people and billions of tons of freight.

Table 4.16

Summary of Pan Europe-Asia Maglev System

- Location of 300 mph Elevated Monorail Maglev Routes
- 3 Routes across Asia – Northern Asia, Middle Asia, and Southern Asia
- 3 Routes across Europe – Northern Europe, Middle Europe and Southern Europe
- 1 route in Middle East
- Connections to Maglev Systems in Other Continents
- Connects to Pan American Maglev System at Bering Strait Crossing
- Connects to Pan African Maglev System at Strait of Gibraltar Crossing and Isthmus of Suez
- 40,000 Total Route Mileage: 30, 000 miles in Trans-Asia Railway Network (TAR) and 10,000 miles for Routes to Bering Strait, in Europe, and in the Middle East.
- Total Investment for 40,000 Miles of the Pan Europe-Asia Maglev Systems is 1.2 Trillion Dollars

## Global Transport of Freight by Maglev Instead of Ships

Today, ships transport an enormous amount of freight. As the Global Maglev Systems begin to operate, this will change. Much of the freight now transported by ship will transition to the Global Maglev Systems.

142

Before describing how Maglev would transport freight that is presently carried on ships, consider first what kinds of freight ships transport, how big is the shipping industry, how rapidly it moves freight, how much energy it uses, how much it costs, and what are its negative aspects and problems.

World shipping is big business. In 2004, it transported 27.6 Trillion ton miles of freight, 10 times greater than the 3 Trillion ton miles of freight transported annually in the US by truck and rail. 6.7 Billion tons are actually shipped, for an average distance of 4,000 miles. (14) Some routes are longer than 4,000 miles, like the 10,000 mile Asia to Europe route.

Figure 4.28
Hanjin Container Ship in San Francisco Bay, Approaching
Golden Gate Bridge

In 2004, there were 46,200 ships in the World shipping fleet, with a total cargo capacity of 548 million dead weight tons (dwt). Of these, 18,150 were general cargo ships, and 1,733 other ships. Ships larger than 80,000 dwt – a major fraction of modern cargo ships – are too big to go through the Panama Canal and must go around Cape Horn.

Ships transport many kinds of cargo, both in bulk form inside the ship – oil, coal, iron ore, etc. – and in closed containers, called TEUs (Twenty Feet Equivalent Unit). TEUs are generally, 20 or 40 feet in length, 8 feet wide, and 8 feet high, with a maximum load capability of 20 short tons (1 short ton = 2,000 pounds). They basically are steel boxes, into which one can load any kind of dry cargo – clothing, electronic equipment, furniture, etc. Large container ships transport thousands of TEUs at a time, as shown in Figure 4.28 for the Hanjin Container ship in San Francisco Bay. The Emma Maersk, one of the World's largest container ships, transports 11,000 TEUs at a time. Total World container ship capacity is 310 million TEUs. Many ports handle 10 million TEUs, or more, per year.

Suppose country A, in Asia, wants to ship some products from the factory or factories where they are produced – or the farms where they are grown – by truck or rail to a main shipping port, where the cargo is put into a TEU, loaded on board a container ship, which sails 10,000 miles to Europe, taking weeks.

After reaching the appropriate port in Europe, the TEU's are unloaded from the container ship and loaded onto trucks or rail cars for transport to their destination. At 1 to 2 cents per ton mile, depending on the shipping spot price, transporting a 20 ton TEU 10,000 miles would cost in the range of 2,000 to 4,000 dollars.

Using the Pan Europe-Asia Maglev System to transport the 20 Ton TEU 6,000 miles from Asia to Europe at 300 mph would have an operating cost of approximately $1,000 per TEU, including propulsion energy cost, vehicle amortization cost, and system personnel costs.

Added to the operating cost is the amortized investment of the guideway. Based on a 30 year amortization of the 180 Billion dollar investment for the 6,000 mile Northern Asia to Europe route and an annual transport of 10 million TEU's, the amortized guideway cost would be 600 dollars per TEU. Total cost would be 1,600 dollars by the Pan Europe-Asia Maglev System, less than by ship, and much faster, 1 day versus 25 days.

And there are substantial additional economic benefits for shipping on Maglev. No need for long distance expensive transport to and from shipping ports – Maglev stations are much more numerous and much more convenient, with fewer loadings and unloading than for ships. With ships, one needs 1 loading to truck or railcar to get it to the seaport, an unloading at the port from the truck or railcar, and a loading onto the ship. Then another 3 unloadings/loadings to get the product to an inland destination on the other side of the Europe-Asia continent, for a total of 6 loadings and unloadings. With Maglev, one loading at a nearby station and one unloading near its final destination.

And the much shorter trip time also results in lower inventory costs, economically more efficient match of production to changes in demand, less overproduction, etc. As the old saying goes, "Time is Money". One day trip-times are much better than 25 day trip-times.

On an energy basis, ships are more efficient than trains or highway trucks. A typical container ship traveling at 26 knots (30 mph) burns 120 gallons of bunker fuel per mile, corresponding to about 750 ton miles of cargo per gallon of oil fuel. (15) Conventional trains transport freight at about 400 ton miles per gallon of oil fuel, and highway trucks achieve approximately 150 ton miles per gallon.

Consists of multiple Maglev vehicles each carrying 40 tons of cargo, traveling at 300 mph, will achieve approximately 400 ton miles per equivalent gallon, using electric energy, not oil fuel. For routes where ships must travel longer distances than Maglev freight vehicles would, the ton miles per gallon for Maglev can comparable to, or even exceed the ton miles per gallon for ships. For example, for the Asia to Europe route, ships travel 10,000 miles, compared to approximately 6,000 miles for Maglev. The shorter distance for Maglev transport makes its energy consumption per ton for the trip comparable to that for ships. And it only takes 1 day, compared to 25 days.

There are substantial negative aspects for ocean ships. First, ships burn more than 7 Billion gallons of oil per day, almost 10% of the World's oil consumption, producing 1.3 Billion metric tonnes of carbon dioxide per year, about 5% of total World emissions. One large container ship emits as much sulfur oxide pollutants as 50 million automobiles do. World ships emit 20 million tons of sulfur dioxide per year, while all of the World's autos emit only 79,000 tons. (16) NOAA (National Oceanic and Atmospheric Administration) estimates that 60,000 people in the World die prematurely from the pollutants and microparticles emitted by the World's ships. (17)

The Global Maglev System will provide fast, efficient, low-cost, transport of passengers and freight across continents and interconnect Europe, Asia, North American, South America and Africa into a World Wide Network. Once implemented, Maglev will transport much of the cargo now carried by ocean ships, with major economic and environmental benefits. Ocean shipping will still be necessary, but the amount will be much smaller, greatly reducing the consumption of fossil fuels and their negative environmental effects.

The investment for the Global Maglev Network is projected to be on the order of 2.7 Trillion dollars, less than 1/1000th of the more than 4,000 Trillion dollars of the sum of the annual World GDP's for the period from 2015 to 2050. The transport savings from the Global Maglev Network will far exceed its investment cost, and the environmental and human benefits will be enormous. The Global Maglev Network will enable safer travel with much fewer accidental deaths and injuries.

# Chapter 5

# Maglev Launch to Space for Power Beaming Back to Earth

## StarTram – The Maglev Launch System

StarTram (1) is a revolutionary new way to launch payloads and people into space using Maglev – much less expensive and with much greater volumes than possible using rockets. StarTram will make it possible to launch solar power systems into orbit that can beam millions of megawatts of clean, very low cost, electric power down to Earth. No need to burn oil, gas or coal to generate power. No need for millions of wind turbines and ground solar power units scattered all over the Earth that only produce power when the wind is blowing or the Sun is shining. Beamed solar power from Space is on call 24/7, whenever you need it, wherever you are on Earth, and whatever amount of power you need.

Since Neil Armstrong stepped onto the Moon on July 20, 1969, we humans have accomplished amazing things in space – Mars rovers, spacecraft journeys to the outer planets, orbiting satellites for communications, GPS, weather observation, and so on. But we've only gone a little way into space and its been very expensive. The Apollo Program to the Moon cost $170 Billion dollars, in today's dollars, (2) and the International Space Station has cost 150 Billion. (3) Using rockets, it costs on the order of $2,000 to $5,000 per pound ($4,000 to $10,000 per kilogram) to launch payloads into Low Earth Orbit (LEO).(4) To place payloads in Geosynchronous Earth Orbit (GEO), 22,000 miles from Earth, the cost is 3 to 4 times greater. To the Moon, Mars, and beyond, much, much greater costs.

With its much lower launch costs, StarTram will carry out 4 critically important missions, not possible with rockets, because of their very high launch costs. The 4 missions and their benefits are:

- Power Earth from space with clean, non-polluting, low cost electric energy, eliminating the need to burn fossil fuels for electricity.
- Protect Earth from asteroid and comet impacts, and reduce ground solar energy input, if necessary, to prevent global temperatures from reaching the danger point.
- Provide a higher quality of life on Earth by more and better communication and observation satellites, and mining scarce resources from the asteroids, moons and planets in the solar system.
- Permit humanity to really explore the solar system. Colonize the Moon and Mars, look for evidence of past or present extraterrestrial life on Europa and other bodies in the solar system that have water/ice, orbit large telescopes to image other star systems in detail, and so on, to better understand humanity's place in the Universe.
- Why will StarTram have much lower launch costs than rockets?

Because it keeps all of its launch equipment on the ground, to be used many times, where rockets throw most of their equipment away when they launch. Think of a company delivering a refrigerator to a customer. It could either deliver it by an expensive helicopter that crashes after delivery, or deliver it by truck, which can be used repeatedly for many deliveries, with its gasoline and driver time the only cost for each delivery. Which choice would the company take? The truck, of course.

Figure 5.1 shows the Saturn V, the rocket that launched Neil Armstrong and his comrades in 1969 to the Moon. 363 feet in height, it was taller than the Statue of Liberty, and weighed 6.5 million pounds. It could deliver 260,000 pounds to LEO, 4 percent of its takeoff weight ended up payload, 6.25 million pounds as trash.

(5) NASA developed the space shuttle (Figure 5.2) to be a reusable launch system to carry cargo and astronauts to the International Space Station (ISS) with the hope that it would reduce launch costs.

Figure 5.1                              Figure 5.2

The shuttle was successful, but still very expensive. At 450 million dollars operating cost per launch, it cost $8,000 per pound to LEO. Including all costs, the whole space shuttle program (it closed down in 2011) cost 196 Billion dollars when adjusted for inflation (6). For the total of 134 Shuttle missions, the average cost per mission was 1.5 Billion dollars.

Besides the very high launch cost of rockets there is the problem of rocket failures. They average 9% -- that is, about 1 out of 10 launches fails. The space shuttle had a lower failure rate, 2 crashes (Challenger and Columbia) out of 133 launches, with all of their crew killed. That's a fatality rate 100,000 times greater than airplanes.

A detailed description of StarTram - Maglev launch system is given in the book, *"StarTram-The New Race to Space"*, by James Powell, George Maise and Charles Pellegrino, available at Amazon.com.

Despite 40 years of intense efforts to reduce the high cost of rocket launch and achieve much greater launch volume, only marginal improvements resulted. The landscape is littered with dead rocket systems – the space shuttle, Titan, Tsyklon, Molniya, START, NASP, Sanger, X-33, X-34, and many others. Rockets appear to have reached their limits, with major advances in the future unlikely.

How does StarTram work? Instead of accelerating spacecraft and their payloads with rocket thrust generated by the combustion of propellants, StarTram uses Maglev. The spacecraft is magnetically levitated and propelled in an evacuated tunnel by the inductive interaction between superconducting loops on the spacecraft and ordinary loops on the tunnel wall. There is no air drag, only the very tiny drag due to electrical resistance losses in the aluminum loops on the tunnel walls.

The cost of the electrical energy required to reach orbital speed in the evacuated tunnel is very small. At a launch velocity of 8 kilometers per second to LEO, the cost of the input electric energy is only $1 per kilogram at 10 cents per kWh. At 10.6 kilometers per second, the launch velocity to reach GEO orbit, the energy cost per kg is $1.80 per kilogram. Compare that with the cost of $5,000 to $10,000 per kilogram for rocket launch to LEO, or the much higher cost for rocket launch to GEO.

After reaching orbital speed in the evacuated tunnel, the StarTram spacecraft exits into the atmosphere through an MHD (Magneto Hydro Dynamic) window that prevents the outside atmosphere from flowing into the evacuated tunnel by MHD forces acting on air ionized by an RF (Radio Frequency) discharge. The spacecraft then coasts upwards to orbit, where an attached small rocket burn finalizes the orbit, preventing the spacecraft from falling back to Earth. For a LEO orbit, the rocket burn is very small,

0.34 km/sec (Figure 5.3). For a GEO orbit the rocket burn is 1.5 km/sec, but still small compared to the 10.6 km/sec required to reach GEO (Figure 5.4).

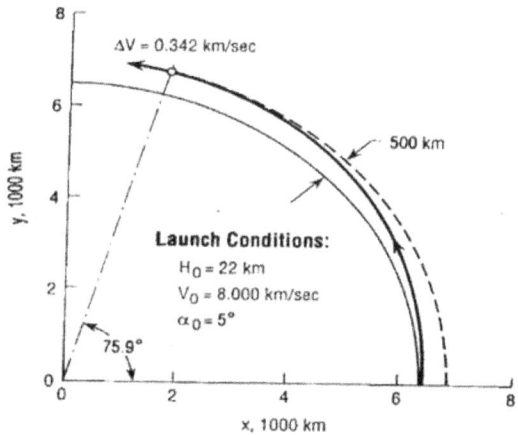

Figure 5.3: Ascent Trajectory to LEO Using StarTram Launch

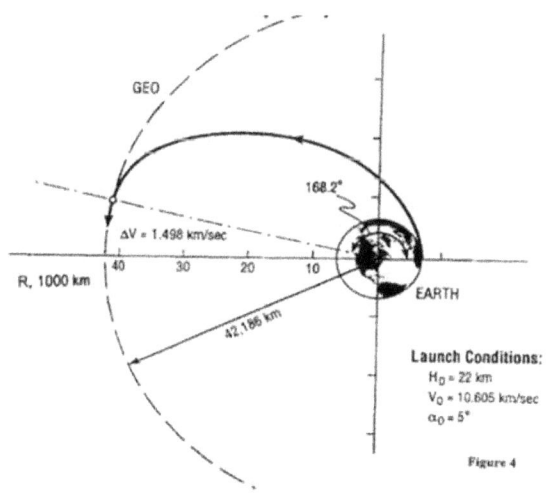

Figure 5.4: Ascent Trajectory to GEO Using StarTram Launch

151

The StarTram spacecraft are magnetically levitated by the magnetic interaction of the superconducting loops on the spacecraft with the ordinary aluminum loops on the wall of the evacuated launch tunnel. The levitation is inherently strongly stable, with magnetic force keeping the spacecraft centered in the tunnel. In addition to the sequence of aluminum levitation loops on the tunnel walls, there is a second set of aluminum loops carrying AC current. The AC current loops magnetically push on the superconducting loops on the spacecraft, accelerating it. Spacecraft speed is controlled by the frequency of the AC current. As the frequency increases, speed increases.

Figure 5.5 shows the layout of the spacecraft loop geometry for levitation and propulsion and how it keeps the spacecraft centered in the tunnel as it accelerates. The StarTram spacecraft uses the same basic Maglev technology that already operates in Japan's 300 mph 1st Generation Maglev passenger system (Figure 5.6), and the proposed 2nd Generation Maglev System (Figure 5.7), described in Chapter 4 on the National Maglev Network and the Global Maglev Network.

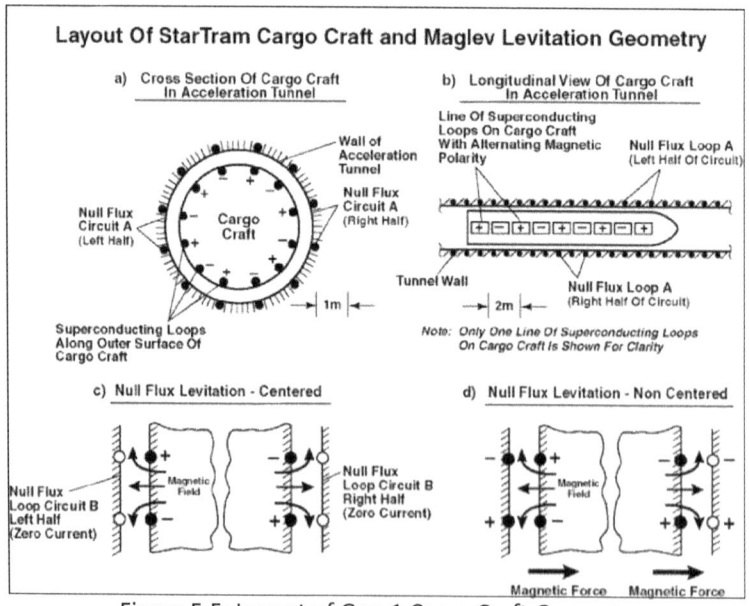

Figure 5.5: Layout of Gen-1 Cargo Craft Geometry.

Figure 5.6

Figure 5.7

153

Two StarTram Systems have been studied. The near term first generation Gen-1 System would launch spacecraft with cargo but not passengers. After reaching orbital speed in the evacuated tunnel, which would be located in high altitude terrain, e.g. at an elevation of 4,000 meters (13,000 feet) or greater, The Gen-1 cargo craft exits from the tunnel at ground level and climbs upward through the remaining low density atmosphere to space (Figure 5.8).

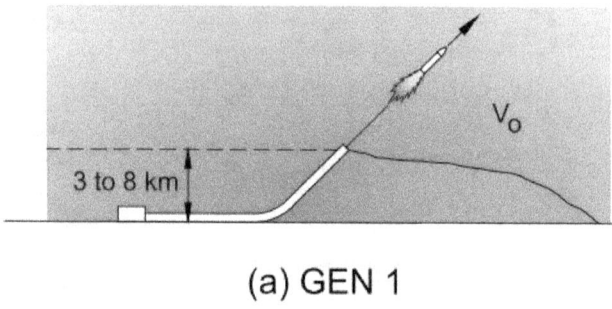

(a) GEN 1

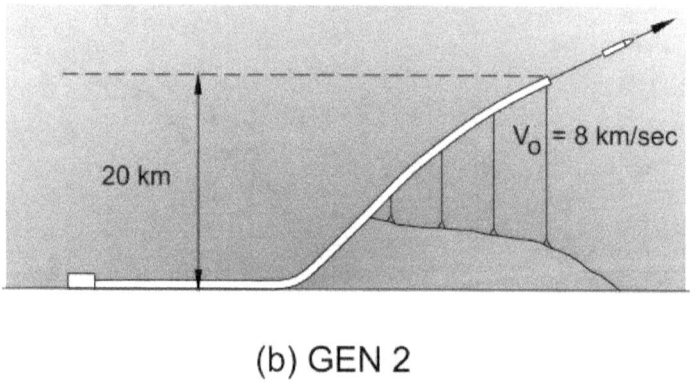

(b) GEN 2

NOTE: NOT TO SCALE

Figure 5.8: Gen-1 and Gen-2 Launch Systems.

The aerodynamic deceleration forces and heating of the Gen-1 cargo craft as it ascends through the atmosphere are strong, but manageable – significantly less than the heating and forces on existing re-entry vehicles for ICBM's (Intercontinental Ballistic Missiles) as they come down from space through the atmosphere to their targets. The re-entry vehicles traverse down through the full atmosphere, including the high density lower portion. The Gen-1 cargo craft can carry much greater amounts of protective coatings and transpiration coolants than is possible for existing re-entry vehicles, which further help them survive their upwards journeys and still manage to survive until they hit their targets.

There are many potential high altitude launch sites around the World for the Gen-1 System – in Alaska, Russia, China, the Andes, Africa, Antarctica, Himalayas, and so on. Depending on site location, Gen-1 cargo craft can be launched into polar or equatorial orbits. Some sites are at very high elevation, like Conga Shan in Szechwan Province in China, where the Gen-1 tunnel exit could be at an altitude of approximately 7,000 meters (23,000 feet).

The atmospheric deceleration forces on the Gen-1 cargo craft are too high, on the order of 8 to 10 g, for passengers. These g forces could be reduced to 2 to 3 g by increasing the mass of the cargo craft and adding a rocket thruster to counter the atmosphere deceleration force. This system, termed Gen-1.5, could transport substantial numbers of persons into orbit.

For the long term, if large numbers of humans are to journey into space, the Gen-2 system may be the way they get there. To avoid atmospheric deceleration forces on the StarTram spacecraft, it transitions from the ground level acceleration tunnel into a magnetically levitated tube (Figure 5.9). Superconducting cables attached to the tube magnetically interacts with a set of superconducting cables located on the ground beneath the tube. Millions of Amperes of current flowing in opposite directions in the 2 sets of cables create a magnetic repulsive force between them that levitates the evacuated tube.

The upper end of the evacuated tube is at an altitude of 20 kilometers (65,000 feet). The StarTram Spacecraft exists through the MHD window (Figure 5.9) into the very low density atmosphere, 5% of the air density at ground level, making the aerodynamic heating and deceleration forces very low.

The Gen-1 StarTram cargo launch system can be built in only 10 years with an aggressive, well-funded development program. It should be a cooperative international program that focuses on important peaceful applications. If carried out separately by nations, it could become a very dangerous and destructive arms race.

Figure 5.9: View of StarTram Exiting the Launch Tube

The longer term Gen-2 StarTram passenger and cargo launch system is much more technically challenging than the Gen-1 system. Gen-1 uses the already existing basic technology for Maglev, MHD, superconducting energy storage, and aerodynamic heating of re-entry vehicles. Gen-2 requires developing magnetically levitated high altitude structures, a whole new R&D area.

Developing it will require much greater funding and take much longer.

Table 5.1 list the basic parameters for the Gen-1 system. One Gen-1 launch facility can launch 10 to 20 cargo craft per day, 330 days per year. The 40 metric tonne cargo craft has a payload of 30 metric tonnes with 10 tonnes for its structure and the small AV rocket that establishes the final orbit. One Gen-1 facility can launch 100,000 metric tonnes or more annually. 100s of times greater than the present total World launch rate. Projected launch cost is 23$ per pound ($50/kg), a 100 times smaller than present rocket launch costs to LEO. Launch cost includes amortization of the StarTram facility, cargo craft cost, and personnel & energy cost.

Table 5.1

| Gen-1 Projected Capital and Operating Costs for Launch to Low Earth Orbit (LEO) | | | |
|---|---|---|---|
| Capital Costs, $B (Nth Facility) | | Operating Cost Per 30 Ton Cargo, $M | |
| Hard rock tunneling (265 km total length, $1500/m3 excavation) | 3.4 B | Cargo Craft Structure (5 metric tonnes, $100/kg) | $0.5 M |
| Aluminum guideway loops ($20/kg) | 0.9 B | Superconductors on craft ($2/KA meter) | $0.43 M |
| SC for energy storage ($2/KA meter) | 0.9 B | Personnel (50 man-days @$500/man day) | $0.02 M |
| Power Conditioning DC to AC for Maglev acceleration ($100/KWe) | 1.0 B | Energy cost (10cents/KWH) | $0.04 M |
| Vacuum & refrigeration systems | 1.0 B | Total Operating Cost Per Launch | $0.99 M |
| Prime power plant [300 MW(e), $3000/KW(e)] | 0.9 B | 10 year amortized capital cost per launch (3650 launches per Year | $0.52 M |
| Buildings and facilities | 2.0 B | Total Cost Per Launch = | $1.5 M |
| Total Capital Cost | 9.1 B | Cost Per kg of Payload = | $50/kg |

The first major mission for StarTram is Powering Earth, by beaming of electrical power down to Earth from space solar power satellites in Geosynchronous Orbit.

Beamed solar power from space would eliminate the need to burn fossil fuels to generate electricity. We already generate non-fossil renewable power from wind and ground solar farms on Earth, but they have problems -- intermittent operation, low average electrical outputs, high cost, negative environmental impacts, and damage from storms and bad weather.

Beamed solar power from space is continuous, 24/7, can readily adjust output to meet peak demands, doesn't harm the environment, cannot be damaged by storms and bad weather, and will be low in cost. Wind power is extremely variable, cannot operate in low or high wind conditions, and its capacity factor is very low, about 30%, delivering power only when wind speed is right. Surface solar farm electrical outputs vary with time of day, cloud cover, etc.

The proposed SPS-ALPHA DRM-5 Space Solar Power System (7) for delivering 2 Gigawatts (e) [2000 Megawatts (e)] has a mass in GEO orbit of 35,000 metric tonnes, equivalent to 17.5 kg/kW (e). The StarTram launch cost to GEO will be greater than to LEO because of the higher launch velocity required to reach GEO orbit. At $100 per kg, the corresponding launch cost per kW(e) of power delivered to Earth would be $100 x 17.5 or about $1,750 per kW (e).

Amortized over 30 years, the launch cost corresponds to about 0.7 cents/kWh, small compared to the 10 cents/kWh for today's electric energy. There are, of course, substantial additional costs for the space power system hardware and the power receiver station on Earth that the power is beamed to.

Using StarTram, launch cost will not be a significant factor in the economic practicality of space power beaming. Using rockets, with a launch cost of $10,000 per kilogram and a mass of 17 kg/kW(e) in GEO, the capital cost of a kilowatt of power from space would be $170,000 per kilowatt, just for launching it – far too expensive to be practical.

One StarTram facility with an annual launch rate of 175,000 tonnes per year could put 10,000 megawatts of beamed space power capacity annually. Over a 10 year period the accumulated beamed power capacity would be 100,000 megawatts. Over 10 years, 5 StarTram launch facilities could put 500,000 megawatts of beamed power into space – enough to meet all of America's present (2013) annual power needs of 4.26 trillion Kilowatt hours (8).

To meet the present (2013) total annual electrical World production of 23.1 Trillion Kilowatt hours (9), 2,600,000 megawatts, a factor of 5 greater than US electrical generation, of beamed electric power would be required. To do this over a 10 year period, 25 StarTram facilities would be needed, each launching 175,000 tonnes per year. Total launch mass over the 10 year period would be 40 million tonnes in GEO orbit – a factor of 13,000 greater than the payload that rockets would launch over a 10 year period.

And that's at present electrical generation rates. The Energy Information Administration (EIA) forecasts an increase to 39 Trillion hours in 2040 AD. (9) And that doesn't include new markets for electrical consumption like electric cars, electrical heating of homes and buildings instead of using fossil fuels, making synthetic fuels for transport, and other applications – even greater launch rates will be required.

Clearly, beamed space solar power will not be possible using rocket launch. Not only would the launch costs be much greater than World GDP, but the billions of tons of propellant exhausts would do unacceptable harm to Earth's environment.

## How Did the idea of Beaming Power from Space Get Started?

In 1968, Dr. Peter Glaser, published in Science magazine (10) his groundbreaking paper on beaming power from Solar Powered Satellites in orbit down to the Earth and filed a patent for it. It took the Patent Office 5 years to finally grant the patent in 1973, probably because it went far outside of the usual Patent Office box.

NASA and ERDA (Energy and Research Agency) started the first serious effort on Glaser's concept in the late 1970's. (11) However, work on the SPS concept (Space Solar Power Satellite) continued in other countries. Then in 1995, NASA got back into the SPS game. In the 20 years since, interest in, and work on, SPS has continued to grow, with studies and experiments in Japan, China, Europe, the U.S. and other countries, as described in Dr. John Mankins book, "The Case for Space Solar Power". (7)

Glaser understood not only how power could be beamed down to Earth from SPS satellites in orbit, but why we would need it. He recognized that Earth's fossil fuel resources were limited and would inevitably run out. He also realized that solar and nuclear energy were the only real options to supply the energy that modern civilization depends on. As discussed in Part I, he was correct and we will need it soon, not only because we will run out of the known reserves in a few decades, but as become evident in the last 20 years, the global warming resulting from fossil fuel emissions of carbon dioxide, will cause environmental catastrophe before the end of the 21st Century.

The basic operational principles of beamed solar power from space are described later. The satellite receives sunlight from the Sun at the same rate as the Earth, about 1300 megawatts (th) per square kilometer. However, if the satellite is in orbit high above the Earth, say in Geosynchronous Orbit (GEO) that is 27,286 miles (35,786 Kilometers) from the Earth at sea level, it sees the Sun virtually continuously.

There is no night in GEO, no sunrise and no sunset. There are no clouds and storms that block out the Sun. As a result, the power generated by a square kilometer of solar collector in GEO will be about 4 times greater than the average power generated by a square kilometer of solar collector located on Earth. Moreover, space solar power beamed down to Earth is continuous. There is no need to store electrical energy to meet demand when the Sun is not shining where one is located. Space solar power beams will continue to provide you with lights at night, and operate your TV, computer, refrigerator, air conditioning, and so on. No need to pay for expensive energy storage in batteries, pumped hydro lakes, etc.

GEO orbits are special. In lower orbits closer to the Earth, spacecraft in their orbits move more rapidly than the Earth rotates, so they don't hover above a particular point on the planet underneath them. In higher orbits above GEO, spacecraft move more slowly in their orbit, than the Earth rotates beneath them. Again, they do not hover above a particular point on the planet. Only in GEO orbit can a spacecraft stay hovered above a particular point on Earth. The ability to stay hovered is extremely important for beamed solar power, since it enables 24/7 power beaming to a given ground receiver – the solar power satellite that beams the power is always in view.

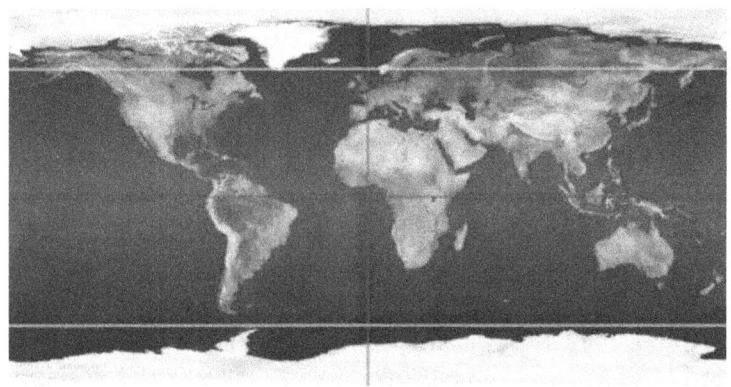

Figure 5.10

There are limitations.

With multiple SPS satellites distributed around the GPS orbit, all of Earth is in view from the SPS satellites, 24/7. However, there are limitations to what locations on Earth are able to receive power from space. Sites with latitudes greater than 60 degrees North of the Equator and 60 degrees South have too great an angle for efficient transmission from the SPS satellite in the equatorial GEO orbit. Fortunately, only a very small fraction of the 7 Billion, World population live below 60 degrees South latitude, as shown in the World map (Figure 5.10).

For the Northern continents, North American, Europe, and Asia, only a tiny part of their countries' population live above 60 degrees North – Alaska in the US and the Eskimos in Northern Canada, in North America. In Europe, only Finland, Sweden, Norway, and Iceland.

In Asia, only Russia, and that in Northern Siberia. All together, total population North of 60 degrees latitude less than 50 million people, compared to the slightly less than 7 Billion that live where SPS could supply their power.

Now to the technology of SPS satellites, described as responses to the questions that people frequently ask:

- How do the SPS satellites generate electrical power?
- How do they beam it down to Earth?
- How do the Earth stations receive the beam and deliver the beamed energy to the electrical grid?
- How large is the SPS satellite and its Photovoltaic Array transmitter?
- How big is the microwave receiver on the Earth?
- Is the microwave beam safe? Will it harm humans, animals and birds that move through it?
- How are the SPS Satellites launched and assembled in orbit?
- How much will beamed space power cost? Can it provide most of Earth's electric power?
- How soon can SPS units be implemented? What are the benefits?
- How does the public get World Leaders to implement Space Solar Power?

**First question**. How do the SPS satellites generate electric power? Very likely SPS will use photovoltaic cells (PV) similar to those already on Earth to generate electric power in large solar cell arrays that supply the electric grid, and on the roof tops of houses that generate power for the home owners.

An alternate approach would be to have reflecting mirrors focus sunlight on a high temperature receiver that used a conventional Rankine or Brayton cycle to generate power. This approach is used by some ground solar power facilities on Earth. Disposing of the waste heat to the atmosphere or water sink from the power cycle for Earth based power plants is relatively easy on Earth. However, it is not simple in space, and the waste heat radiators are heavy and complex. Thus, it is very likely the SPS satellites will use PV cells to generate their electric power.

SPS PV arrays will have a number of important advantages over ground solar power PV arrays located on Earth.

- Much greater electric output per unit area, compared to arrays on Earth,
- Not subject to damage by storms and bad weather
- Able to serve multiple areas widely distributed across the planet without needing transmission lines
- Very low environmental impact on an array of PVs
- Continuous power, not intermittent

At 30% efficiency, sunlight to DC electricity, an array of one square kilometer of PV cells will generate approximately 400 megawatts(e) of electric power. The same array on Earth would generate only 1/4th as much power, i.e. 100 megawatts (e).

Besides having much more power per unit area of the solar PV Array, the other advantages of SPS are also very important. Space based solar PV arrays cannot be damaged by storms, can transmit continuous power and meet peak demands at many different locations, without needing transmission links – things Earth based arrays are not capable of.

**Second question**. How do the SPS satellites beam power down to Earth? The preferred approach is to convert the DC electric power from the PV arrays to microwave power that is focused by a large transmitter and beamed down to receivers on Earth. Optimum microwave frequency is in the range from 5 to 10 gigahertz. In this frequency range, atmospheric attenuation of the microwave beam is very small, less than 1%. (2) Including focusing lenses, the overall beaming efficiency is in the range of 92% to 94% of the microwave power put into the beam ends up as microwave power at the receiver on Earth.

Based on a conservative conversion efficiency of 70% from DC to microwave power (11), 1,000 megawatts of DC electric power generated by the PV array, the microwave power level reaching receivers on Earth is on the order of 0.70 x 0.93 x 1,000, equaling 650 megawatts. With increased DC to microwave efficiency, which appears possible, the microwave power level at receiving stations would be greater.

**Third question.** How do the Earth Stations receive the beam and deliver the beamed energy to the electrical grid? The ground based rectenna (Figure 5.11) receives the microwave beam, and converts it to ordinary 60 Hertz AC current that is fed to the electrical grid that it is connected to. The conversion of efficiency, microwave to AC power is on the order of 85% (11).

Figure 5.11
Space to Ground Microwave, Laser Pilot Beam

Putting the individual efficiencies together, the overall end-to-end efficiency, DC power in space to AC power in electrical grids on Earth, is then:

End-to-end efficiency = 0.70 x 0.93 x 0.85 = 0.55. That is, for every 1,000 MW(e) generated by PV arrays in space, about 550 MW(e) ends up power in electricity grids.

**Fourth and Fifth questions**. How big is the satellite PV array and its transmitter and the microwave receiver on Earth? The two questions are tied together by the physics of focusing microwaves. The diameters of the receiver and transmitter are related, plus the wavelength of the microwave beam and the distance between transmitter and receiver are related by the equation below. (7)

$D_{TRANS}$ x $D_{RCVR}$ = 2.44 $\lambda_{BEAM}$ x $R_{Distance\ Trans\ to\ RCVR}$

For an SPS Satellite in GEO orbit, R = 3.58 x $10^7$ meters

For a 10 gigahertz microwave beam

$\lambda_{BEAM}$ = 3x $10^{-2}$ meters

Incorporating R & $\lambda$, equation (1) is then

$D_{TRANS}$ x $D_{RCVR}$=2.44x3x$10^{-2}$x3.58x$10^7$=2.62x$10^6$m$^2$

Expressed in kilometers

$D_{TRANS}$ x $D_{RCVR}$ = 2.62 km$^2$

If we make $D_{TRANS}$ the same size as $D_{RCVR}$, then

$D_{TRANS}$ x $D_{RCVR}$ = 2.62 = 1.62 km or about 1 mile in diameter. If we make $D_{TRANS}$ bigger. E.g., with $D_{TRANS}$ = 2 kilometers, $D_{RCVR}$ is smaller= 1.3 kilometers

The sizes of the transmitter and receiver are independent of the amount of power that is transmitted, so it pays to transmit a lot of power per transmitter/receiver combinations. For an AC power of output of 2,000 megawatts(e) at a receiving station on Earth, one would have to generate 2,000/0.55 = 3,600 megawatts(e) of DC power in space, based on an end-to-end efficiency of 55%. The total area of the solar cell arrays to provide 3,600 megawatts(e) would be

9 km², based on 400 megawatts(e) per km² of area. Multiple PV arrays, each with a smaller area, could be connected to the transmitter.

**Sixth Question**, Is the microwave beam safe? Will it harm humans, animals, and birds that move through it? Based on studies of flora and fauna exposed to microwave, there appear to be no harmful effect. To assure safety the International Academy of Astronautics (IAA) has recommended in its 2008-2011 study of space solar power that the maximum intensity of wireless power transmission should be less than full summer sunlight at the equator, i.e., 1,000 watts/m².(7) A 2 km diameter receiver receiving 2,350 megawatts of microwave power and delivering 2,350 x 0.85 = 2,000 megawatts of electric power to the grid, would have a microwave beam intensity at the receiver of:

$2,350 \times 10^6/(\pi/4)(2,000)^2 = 748$ watts/m², well under the IAA recommended limit. Microwave intensity outside of the receiver, which would be fenced off, would be much smaller.

Figure 5.12

Figure 5.13

**Seventh question.** How is the SPS system launched and assembled in orbit? Many different designs and assembly approaches have been proposed. Figures 5.12 and 5.13 show 2 of the proposed designs. The Sun Tower approach (Figure 5.12) is a long column of modular units with the microwave transmitter at the bottom of the column, pointed towards Earth. The column is gravity stabilized, so the transmitter always remains at the bottom, facing the receiver on the surface of the Earth, to which it radiates power.

The modular Sandwich approach is shown in Figure 5.13. This design reduces the area of the PV array that generates DC power by using a set of mirrors to concentrate sunlight on the PV cells at higher intensity than the ambient 1,300 watts per m². Mankins (7) describes in detail the wide variety of SPS designs that have been proposed. In general, these designs require human or robotic assembly of a large number of modules that have been launched into space by rockets.

An alternate approach is based on magnetically inflated cable (MIC) structures as shown in Figures 5.14 and 5.15. Launched as a compact collapsed package of super conducting cables, tethers, and thin PV films, when the superconducting cables are energized with current, the resultant magnetic forces automatically expand the collapsed package into its final full size structure.

## THIN FILM SOLAR CELL SYSTEM

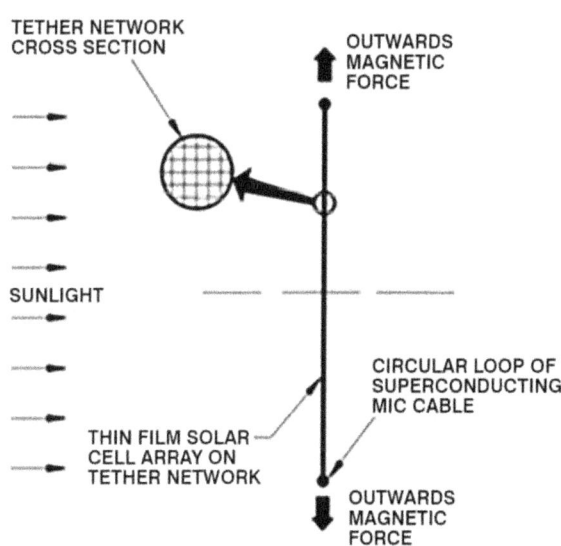

Figure 5.14

Two versions of a MIC solar PV array are shown in Figures 5.14 and 5.15. The first (Figure 5.14) is a simple loop of superconducting cable, with tethers of Kevlar or other high strength fibers that restrain the outwards magnetic forces acting on the cable loop.

## SOLAR CONCENTRATOR SYSTEM

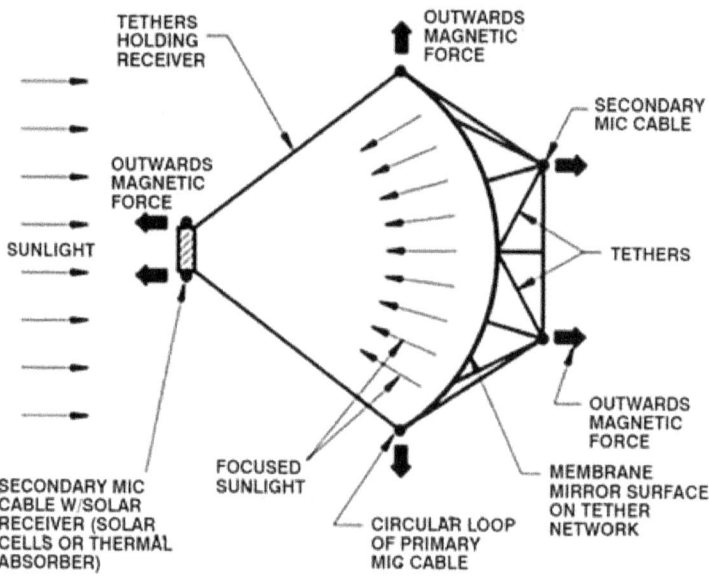

Figure 5.15

The second version (Figure 5.15) uses superconducting cables to form a large thin-film reflecting mirror that concentrates sunlight onto a small solar cell PV array, reducing its mass and cost. A concentration ratio in the range of 5 to 10 can be achieved. For a 2 mil (50 microns) thick reflecting film of aluminized plastic, the mass of the solar collector would be less than 0.2 kg/kw(e), very small compared to other SPS designs.

A third and even simpler version of a MIC space power satellite is a small flat rotating MIC loop structure with extended fins that hold the PV cells. The extended fins would extend beyond the disc to a distance of 3 to 4 times than the radius of the MIC loop, with the outwards centrifugal force from the rotation keeping them as flat sheets.

For an effective fin thickness of 100 microns, structure plus PV cells, the unit weight of the fins would be only 0.3 kg/kw(e) at a solar cell efficiency of 30%. The mass of the rotating MIC loop would be much less than the mass of the fins.

The solar PV array is just part of the total SPS mass in orbit. To assess total SPS mass and its cost, including launch cost, we use Mankins Design Reference Mission (DRM) #5 for the SPS-ALPHA concept described in his book, The Case for Space Solar Power (7). Mankins describes the SPS-ALPHA concept in detail – its various components, their mass and cost, how they are assembled in orbit, and so on. Five DRM versions of SPS-ALPHA are described, with increasing levels of beamed power to Earth. DRM#1 and DRM#2 are small scale demonstration SPS systems in LEO orbit. DRM #3 is a subscale integrated demonstration in GEO orbit. Two beamed power levels for DRM#3 are analyzed – one at 2 megawatts (e) and the other at 18 megawatts (e). DRM#4 is a commercial SPS system delivering 500 megawatts (e) to markets on Earth.

Table 5.2
Parameters for 2000 MW SPS-ALPHA Space Power Systems
Basis – DRM Case #5

| Parameter | Value |
|---|---|
| Beamed Power Input Electrical Grid on Earth Per Unit | 2,000 megawatts(e) |
| Location in Space | GEO orbit |
| SPS Platform Hardware Mass in Orbit per Unit | 34,800 tonnes |
| DRM#5 Hardware Cost Per Unit | $5.7 Billion |
| Earth Receiver Cost Per Unit | $0.7 Billion |
| Total Hardware Cost (Launch Cost Not Included) | $6.4 Billion |

DRM#5, the design we have selected, would deliver 2,000 megawatts(e) [2 gigawatts(e)] of power to markets on Earth. Table 5.2 summarizes the principal parameter for DRM#5.

**Eighth Question** – how much will beamed space power cost? Can it provide most of Earth's electric power? Based on Mankin's cost projections (7) for the SPS ALPHA-DRM#5 design and a StarTram launch cost of $100 per kilogram to GEO orbit, the parameters for the SPS units launched by StarTram are shown in the following Table 5.3.

Table 5.3
Parameters for StarTram Launch Facility to Place 2,000 MW(e) SPS ALPHA-DRM# Units in GEO

- 40 metric tonne payload per launch
- 12 launches per day
- 175,000 metric tonnes launched per year
- 5 SPS ALPHA – DRM#5 Units emplaced in GEO per year (2,000 MW(e) per unit, 34,800 tonnes per unit.
- 30 Billion dollars Capital cost of StarTram facility
- $100/kg launch cost per kg to GEO (amortized)
- 3.5 Billion dollars cost to launch 1 SPS ALPHA-DRM #5 Unit [$1250/kw(e)]
- First StarTram launches begin in 2025
- 2 potential rates of StarTram Facility Construction Studied:
- Case 1: 2 facilities constructed per year in years 2025, 2026, 2027, 2028, and 2029 10 total Star Tram facilities constructed. Total Construction Cost of $300 Billion Dollars.
- Case 2: 4 facilities constructed per year in years 2025, 2026, 2027, 2028, 2029. 20 total StarTram Facilities constructed. Total construction cost of $600 Billion dollars.
- 10 Billion dollars total capital cost for one 2000 MW(e) SPS Unit Space and Receiver Hardware Plus Launch Cost.
- 2 cents per KWH cost of SPS power beamed to Earth (Amortized Capital plus O&M costs)

The StarTram launch cost of $100 per kilogram to GEO is 2 times the $50 per kilogram to LEO, reflecting the high launch speed for a direct launch to GEO, higher delta velocity (1.5 km/sec), and heavier rocket required to establish the final orbit in GEO. An alternative approach is to launch to a lower orbit, e.g. LEO, and use a small high Isp electric thruster powered by the solar PV array to slowly ascend to GEO orbit. Detailed analysis will be needed to determine which approach is best.

Using StarTram, launch cost is approximately 3.5 Billion dollars for the 2,000 MW(e) SPS systems, about 1/2 of the systems hardware cost of 6.4 Billion dollars, resulting in a total cost of approximately 10 Billion dollars, equivalent to $5,000/kW(e). If present rocket launch costs to GEO, on the order of $10,000 per kilogram were to apply, the total capital cost would be 360 Billion dollars for the 2,000 MW(e) SPS System, equivalent to $180,000/kW(e), 36 times greater than launching by StarTram. A completely impractical cost.

Using StarTram, the 2000 MW(e) SPS system can deliver 1.75 x $10^{10}$ KWH/year of electric power at a cost of 2 cents per KWH. The US consumed 4.3 Trillion KWH of electric power annually in 2013.(8) 245 SPS units in space could supply all of America's 2013 electric power generation. At 34,000 tonnes per unit, 1 StarTram launch facility with a launch capability of 175,000 tonnes per year could launch 5 each 2,000 MW (e) SPS units annually and 75 units over a 15 year period. 5 StarTram facilities could launch the 245 units in slightly less than 10 years. The World consumed approximately 20 Trillion KWH of electrical power in 2011, about 5 times the annual US power consumed in 2011 (9). A correspondingly greater number of 2,000 MW(e) SPS units would be required, about 1140.

But, World electric power needs will grow substantially in the decades ahead. The EIA forecasts that World electric power generation will increase from 20 Trillion KWH per year in 2011 to 39 Trillion KWH/year in 2040, a factor of 2 increase to supply all of the World's electrical power needs. Extrapolating the 2040 EIA projection to 2050, World electric generation will further increase at approximately 0.6 Trillion KWH/year for a total of 45 Trillion KWH/year. To supply all of the World's electrical power in 2050 AD would require 2,400 SPS units in orbit (Table 5.5). And, as discussed in Chapter 7, even more electric power will be required if synthetic gasoline and diesel fuel for motor vehicles and jet fuel for airplanes are manufactured from carbon dioxide extracted from the atmosphere.

Yes, beamed solar power could supply virtually all of Earth's future electric power needs, and supply it indefinitely with no environmental damage and no additions of greenhouse gases to the atmosphere.

And, it can accomplish the above, while at the same time, saving enormous amounts of money for electrical consumers, because of its much lower cost per KWH. SPS power to the grid will cost about 2 cents per KWH – even less as SPS designs evolve and become cheaper and lighter – while commercial power from conventional fossil fuel plants now costs about 10 cents per KWH(e), and will become even more expensive in the years ahead. (12)

The capital cost to put 2400 SPS units in orbit to supply all of the World's power in 2050 AD, is 24.6 Trillion dollars including the cost of constructing 20 StarTram launch facilities.(Table 5.6) The Net World Savings from SPS power at 2 cents per KWH, as compared to purchasing conventional power at 10 cents per KWH is 40.6 Trillion dollars over the period from 2025 to 2050AD (Table 5.4). The savings in the cost of power is almost 2 times the cost of putting the SPS units in orbit. And, as noted above, it is likely that SPS units will become even cheaper as the technology evolves, further increasing the savings.

World GWP is presently 65 Trillion dollars annually and probably will increase substantially in the coming decades. Projections vary as to what the increases will be for the various nations of the World. Using the predictions of Price Waterhouse

Coopers (13) for the 20 largest economies in the World the following are the top 10 ranked countries in 2014 and 2050AD.

China's GDP grew by a factor of 5 from 10.0 in 2014 to 50.9 in 2050. India grew by a factor of 10.5 from 2.0 in 2014 to 21.0 in 2050. The 2 most populous countries on Earth, with approximately 1/3 of the total World population in 2050. European countries – France, Germany, and United Kingdom – grew only by a factor of about 2. Italy dropped out of the top 10, also growing only be a factor of 2. Japan grew by only a factor of 1.6 and the U.S. grew by a little more than a factor of 2. Russian and Brazil grew by a factor of 3.

Table 5.4

| | 2014 Annual GDP | | | 2050 Annual GDP | |
|---|---|---|---|---|---|
| | Country | GDP (in 2014 Trillions USD) | | Country | GDP (in 2014 Trillions USD) |
| 1. | U.S. | 17.5 | 1. | China | 50.9 |
| 2. | China | 10.0 | 2. | U.S. | 38.0 |
| 3. | Japan | 4.8 | 3. | India | 21.0 |
| 4. | Germany | 3.9 | 4. | Japan | 7.8 |
| 5. | France | 2.9 | 5. | Brazil | 7.4 |
| 6. | United Kingdom | 2.8 | 6. | Russia | 6.2 |
| 7. | Brazil | 2.2 | 7. | Germany | 6.2 |
| 8. | Italy | 2.2 | 8. | United Kingdom | 5.9 |
| 9. | Russia | 2.1 | 9. | Mexico | 5.8 |
| 10. | India | 2.0 | 10. | France | 5.7 |

China, the US, and India will be the dominant economic powers in 2050, with their combined GDPs being almost 60 percent of total annual World GDP, which will be on the order of 200 Trillion dollars, 3 times greater than today's (2014) annual World GDP.

And many other developing nations in Africa, Asia, and South America will want to catch up to the big GDP nations, so 2050 is not a final state of affairs. Given low cost, environmentally clean SPS power, they will seek to use it to develop their economies and a higher standard of living, so that electrical power demand will still continue to substantially increase over the remaining 2050 to 2100 period of the 21st Century.

The 10-year Phase 1 program would develop and test the components of the StarTram launch system and the SPS system to be ready for commercial deployment in 2025. Here we examine only the development and testing of the StarTram System – planning the program for the development and testing of the SPS system will be carried out by other scientists and engineers.

**Ninth Question**. Based on our studies of StarTram technology and its development requirements, we believe that StarTram launch facilities can begin operation in 2025. We assume that the SPS technology will also be ready for implementation in 2025. However, this will have to be validated by R&D on the SPS system. If SPS technology is not ready in 2025, the implementation schedule will slip to a later date. We believe that 2025 target date for the SPS system should be feasible, but if it is delayed to 2030 or 2035, SPS power can still meet a major portion of the World's power needs in the 21st Century.

- Based on a 2025 start date for SPS launches, implementation schedules have been analyzed (Table 5.5) for Phase 2.
- Case 1: 10 Total StarTram launch facilities constructed in years 2025 through 2029, 2 per year.
- Case 2: 20 total StarTram launch facilities constructed in years 2025 through 2029, 4 per year.

As each StarTram launch facility is completed and begins operation, each year it launches 5 SPS units into GEO orbit. Table 5.6 gives the total SPS units in orbit at the end of year 2029. For Case 1, 10 total StarTram launch facilities, there are 150 SPS units in orbit. For Case 2, 20 total StarTram launch facilities, there are 300 SPS units in orbit.

Already in 2029, the SPS units supply a significant fraction of the total World electric power generation of 32.1 Trillion KWH. For Case 1, SPS supplies 2.7 Trillion KWH (Table 5.5), 8.4 percent of the World Total for Case 2, SPS supplies 5.40 Trillion KWH (16.8 percent).

Going forward from 2030 in Phase 3 the already operating StarTram launch facilities continue to launch more SPS units, increasing the amount of SPS power beamed to Earth. (Table 5.5). In 2050 AD, Case 1 has 1260 SPS units in orbit, supply 48% of total World electric generation while Case 2 has 2400 SPS units in orbit supplying 76% of total World generation.

The corresponding annual investments for the SPS units, together with the annual savings in World electric power cost, based on 2 cents /KWH average cost for power from conventional sources (8) – coal, gas, nuclear, wind, and ground solar PV – are shown in Table 5.6.

By 2035 the net annual savings offset the net annual investment in SPS units. From then on the net annual savings are greater than the annual investment. By 2050, for Case 1 the total SPS investment over the period 2025 is 12.30 Trillion dollars, while the total savings in power cost is 20.28 Trillion dollars, resulting in a net savings for the period of 20.28-12.30 = 7.98 Trillion dollars. By 2050, for Case 2, total investment is 24.60 Trillion Dollars, total savings are 40.56 Trillion dollars, and net total savings are 15.96 Trillion dollars.

Total annual GDP in 2050 for the top 20 largest economies in the World is projected to be 188 Trillion dollars, expressed in 2014 US dollars (13). China's annual GDP is 50.8 Trillion USD, America is 38.0 Trillion USD, India, 20.8 Trillion USD, accounting for more than ½ of the 190 Trillion US dollars total. The other 17 economies, Japan, Russia, Brazil, Germany, etc., down to #20, Switzerland at 2.0 Trillion dollars annually, make up the balance. It is noteworthy that China will have a bigger GDP than America, and that India will have a DDP that is more than 1/2 that of the US.

In the years following 2050 Case 2 will have a net annual savings in power cost of about 45 Trillion KWH x $0.08/KWH=3.6 Trillion dollars, about 2% of World Annual GDP. Not included in the savings enabled by the SPS-StarTram programs are the much greater external costs of generating electric power from fossil fuels – the costs of global warming from greenhouse gas emissions, strip mining, oil spills, pollutants that damage health, contamination of aquifers from fracking for gas and oil, and so on. The benefits from eliminating the external costs are even greater than the economic benefits.

**Tenth Question.** How does the Public get World leaders to implement Space Solar Power? This is the most difficult question of all. It's easy to make a very strong case for beaming Space Solar Power to Earth as a replacement for electric power generated from fossil fuels. There are large economic and environmental benefits, humanity can avoid the real possibility of the collapse of modern civilization when we run out of fossil fuels and/or environmental catastrophe occurs.

In the absence of opposition from vested powerful economic interests, there would be strong political support from World leaders, who would initiate an aggressive program in SPS because of its many major benefits to their constituents.

Unfortunately, political leaders also listen carefully to powerful existing economic interests that would be negatively impacted by implementing SPS. These include the fossil fuel industries that produce oil, gas, and coal, electric utility companies, industries that manufacture conventional equipment – turbines, boilers, etc. – for generating electric power, and so on.

To counter this opposition, it will be necessary for the public and environmental groups to recognize the benefits of SPS and strongly push for its implementation.

In summary, the StarTram SPS project is extremely important. It offers the opportunity to meet World electrical power needs without using polluting, environmentally damaging fossil fuels, at lower cost than conventional power. It can provide sustainable electric power for humanity indefinitely. Future generations will not have to struggle for existence in an increasingly environmentally devastated World as fossil fuels run out.

Table 5.5
Implementation Schedule for SPS ALPHA DRM#5 Units in Orbit and SPS Power Beamed to Earth.
Basis:  5 SPS Launches Per Year from 1 Star Tram Facility
2,000 MW(e) per SPS Unit
EIA Projections for World Power Demand ( )

| Year | Number of StarTram Facilities | | SPS Launches per year | | Number of SPS Units In Orbit | | SPS Power Supply Trillion KWH/year | | Total World Power Demand |
|---|---|---|---|---|---|---|---|---|---|
| | Case 1 | Case 2 | Case 1 | Case 2 | Case 1 | Case 2 | Case 1 | Case 2 | Trillion KWH/YR |
| 2025 | 2 | 4 | 10 | 20 | 10 | 20 | 0.18 | 0.36 | 29.5 |
| 2026 | 4 | 8 | 20 | 40 | 30 | 60 | 0.54 | 1.08 | 30.1 |
| 2027 | 6 | 12 | 30 | 60 | 60 | 120 | 1.08 | 2.16 | 30.8 |
| 2028 | 8 | 16 | 40 | 80 | 100 | 200 | 1.80 | 3.60 | 31.4 |
| 2029 | 10 | 20 | 50 | 100 | 150 | 300 | 2.70 | 5.40 | 32.1 |
| 2030 | 10 | 20 | 50 | 100 | 200 | 400 | 3.60 | 7.20 | 32.7 |
| 2035 | 10 | 20 | 50 | 100 | 450 | 900 | 8.1 | 16.2 | 35.8 |
| 2040 | 10 | 20 | 50 | 100 | 700 | 1400 | 12.6 | 25.2 | 39.0 |
| 2045 | 10 | 20 | 50 | 100 | 950 | 1900 | 17.1 | 34.2 | 42.0 |
| 2050 | 10 | 20 | 50 | 100 | 1200 | 2400 | 21.6 | 43.2 | 45.0 |

Case 1:  10 StarTram Launch Facilities – SPS supplies 48% of Total World Power Demand

Case 2:  20 StarTram Launch Facilities – SPS Supplies 96% of Total World Power Demand

Table 5.6
Investment In and Savings in Cost of Beamed World Power from SPS ALPHA – DRM# solar Satellites in GEO Orbit

| Year/ Period | Number of SPS Units in Orbit | | SPS Investment During Year/Period In Trillion Dollars | | | SPS Power Savings During Year/Period in Trillion Dollars | |
|---|---|---|---|---|---|---|---|
| | Case 1 | Case 2 | Case 1 | Case 2 | | Case 1 | Case 2 |
| 2025 | 10 | 20 | 0.16 | 0.32 | | 0.01 | 0.02 |
| 2026 | 30 | 60 | 0.26 | 0.52 | | 0.04 | 0.08 |
| 2027 | 60 | 120 | 0.36 | 0.72 | | 0.09 | 0.18 |
| 2028 | 90 | 180 | 0.46 | 0.92 | | 0.14 | 0.28 |
| 2029 | 150 | 300 | 0.56 | 1.12 | | 0.22 | 0.44 |
| 2030-2034 | 400 | 800 | 2.50 | 5.00 | | 2.16 | 4.32 |
| 2035-2039 | 650 | 1350 | 2.50 | 5.00 | | 3.96 | 7.92 |
| 2040-2044 | 900 | 1800 | 2.50 | 5.00 | | 5.76 | 11.52 |
| 2045-2050 | 1200 | 2400 | 3.00 | 6.00 | | 7.90 | 19.80 |
| Total SPS Investment in Trillions $ | | | 12.30 | 24.60 | Total SPS Power Savings in Trillions $ | 20.28 | 40.56 |

Net Savings From SPS System=Total SPS Power Savings – Total SPS Investment

Case 1: Net Savings = 20.38 T$ - 12.30 T$ = 7.98 Trillion $

Case 2: Net Savings = 40.56 T$ - 24.60T$ = 15.96 Trillion $

178

# Chapter 6

# Maglev Storage of Energy from Wind and Ground Solar Sources

The US electrical grid provides reliable electrical energy to virtually every person, industrial, and commercial entity in America on demand, usually without problems, at acceptable cost. Long distance transmission lines interconnect the many large power generation plants around the United States and Canada, to the millions of users scattered across the country, transmitting many thousands of megawatts over distances of many hundreds of miles.

However, there are limitations and constraints for the US electrical grid. Electrical demand varies considerably and often rapidly with the time of day, day of week, and season. Baseload generating plants cannot meet rapidly shifting demands, because their time response is too slow. To react to fluctuating demands, quick response peaking power plants, usually fueled by natural gas, are used. About 20% of the annual kilowatt hours (kWh) consumed in the US comes from peaking power plants. The cost of peaking power is substantially greater than the cost of power from baseload power plants.

If baseload power plants could store their electrical output at low cost, they could completely eliminate the need for peaking power plants. They would generate surplus power at night, store it at a few cents per kilowatt hour and feed it back to the grid during peak power periods instead of having to generate electrical power at much greater cost using peaking power plants.

That is now. The situation will become much more challenging if the US begins to shift away from fossil fuel power plants towards renewable wind and ground solar power sources. If beamed space power is developed as described in Chapter 5, it will become the principal source of electric power for the World. If it is not developed, however, the World will have to rely on wind and ground solar power sources. The outputs from wind and ground solar are highly variable and non-predictable. When the wind blows – which is typically only 30% of the time – the generated power may or may not match demand. Sometimes there will not be demand for the wind power; other times there will be demand, but the wind isn't blowing. Ground solar power also fluctuates with time of day, clouds, and seasons. The ground solar power output curve typically does not match the grid demand curve.

Assuming that beamed space solar power (Chapter 5) is not developed, unless the power outputs from wind and ground solar power facilities can be stored efficiently and low cost in large amounts at many locations in the US, to be fed to the grid when customers demand power, wind and ground solar will not become significant sources for US energy needs. America would then have to continue to primarily rely on fossil fuel power. The carbon dioxide emissions from fossil fuel power plants would continue, and global warming would get worse.

Matching increasing variable output to demand in the years ahead will require new energy storage technologies. Pumped hydro storage is practical but has siting restrictions and major environmental concerns. Total US electrical generation capacity is 1 million megawatts, while the pumped hydro capacity is only 22,000 megawatts – about 2% -- and it's unlikely to get much larger. (1) Moreover, potential wind and ground solar power sources are often far distant from any possible pumped hydro storage type facility. Also, the output/input energy efficiency of pumped hydro is only about 70%, so a substantial fraction of the input power is lost. (2) Compressed air storage in underground cavities is a recent technology development, but it too has limitations and concerns. One requires stable geologic conditions, and output/input power losses are substantial.

Pumped hydro and compressed air storage are in the category of macro, i.e., bulk energy storage, with the capability of storing many hundreds of megawatt hours at high, but still acceptable cost.

The other energy storage options – flywheels, batteries, hydrogen fuel cells, thermal, and superconducting magnetic energy storage (SMES) – fall into the category of micro energy storage. They are too expensive and too limited in capacity to store large amounts of electrical energy. They can play a very useful dynamic role, however, in helping to ensure that the various sections of the grid system that connect together have uniform AC frequency and phase.

In addition to meeting time varying power demands and ensuring proper frequency and phasing in the grid, energy storage systems, particularly bulk storage systems, can provide insurance against grid failures due to accidents and sabotage. Several years ago, some trees fell during a storm in Ohio, disrupting electrical transmission capability. As a consequence of propagating failures, the US Northeast was blacked out for a substantial length of time. Sabotage is an even greater worry. It can be physical in nature – blowing up substations and/or transmission lines, attacking power generation plants, etc. – or it can be cyber in nature – shutting down power plant controls, substation switches, etc.

Large scale, low cost bulk energy storage would greatly help ensure that accidents or sabotage would not shut down large sections of the US electrical grid, by providing electrical energy for extended periods of time to make up for the portions that were temporarily not working.

The new Maglev energy storage technology, MAPS (Maglev Power Storage), can:

- Store energy from baseload power plants to satisfy peak power demands at low cost
- Store energy from highly variable wind and ground ground solar power sources
- Store energy to maintain the electrical grid in the event of sabotage and accidents

The advantages of MAPS include:

- Very high storage efficiency – over 90% of the input electrical energy to a MAPS storage facility can be returned back to the electrical grid
- Ability to store large amounts of electrical energy at low cost, only 2 to 3 cents per kWh
- Ability to be built for a wide range of electrical storage capability from thousands of kilowatt hours to thousands of megawatt hours.
- Ability to be sited at a very wide range of locations all over the US
- Much less environmental problems than pumped hydro.

## The US Electrical Grid – Current Status

Table 6.1 shows the installed US electrical generation capacity by type of generation, plus the amounts of electricity actually generated from the various sources. Figure 6.1 shows the actual amounts expressed as a fraction of the total actual generation.

While natural gas generation capacity 455,000 megawatts is considerably greater than coal-fired generation capacity (360,000 megawatts), the actual amount of electrical energy generated is considerably less (883 million megawatt hours) than that for coal plants (1,996 million Megawatt Hours). Figure 6.2 shows the capacity factors for the fraction of time that the plant actually operated for the various types of generation. (4)   Nuclear was highest, 91%, followed by 72% for coal, and 47% for natural gas.

Because of the highly variable output and low capacity factor of wind power – the wind speed is only strong enough to generate power about 30% of the time – it is likely to remain at the few percent level unless a new large-scale, efficient, low cost energy storage system is developed. The present energy storage systems cannot meet the requirement needed for large scale implementation of wind and ground solar power sources.

Wind electric power generation is only a very small fraction of US and World generation, being about 1% of US total generation and 2% of World generation. The EIA (Energy Information Administration) projection is that World generation from wind power will slowly climb to almost 5% of the World total by 2035 – a lot less than wind power advocates want and expect.

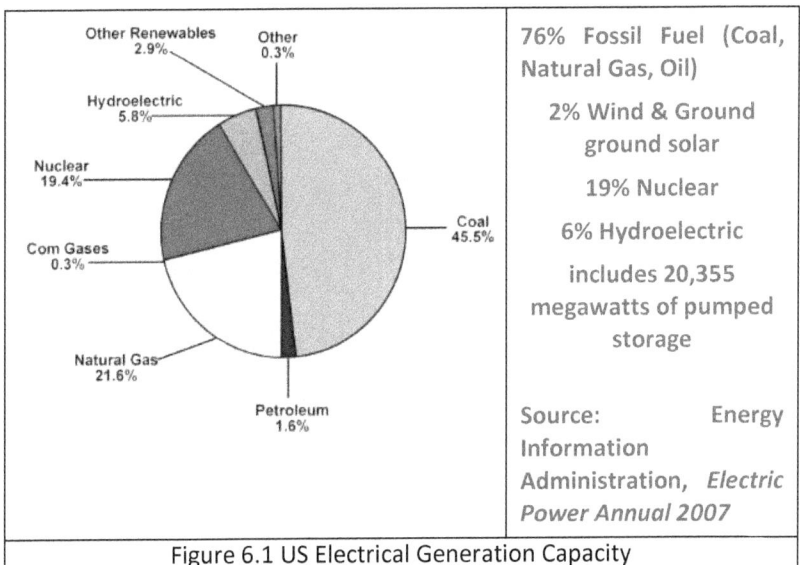

76% Fossil Fuel (Coal, Natural Gas, Oil)

2% Wind & Ground ground solar

19% Nuclear

6% Hydroelectric includes 20,355 megawatts of pumped storage

Source: Energy Information Administration, *Electric Power Annual 2007*

Figure 6.1 US Electrical Generation Capacity

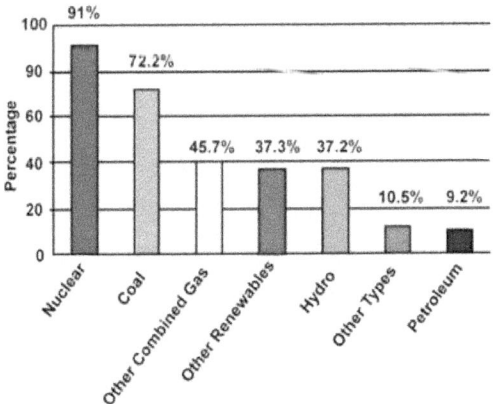

Figure 6.2 Average Capacity Factor By Source

183

## Table 6.1

Megawatts of Installed Electric Capacity and Megawatt Hours of Actual Electric Energy Generation Per Year for the United States as a Function of Generation Type.
### Basis:  EIA 2008 Data (1,3)

| Type Generation | Installed Capacity (Megawatts) | Annual Electric Generation (Million Megawatt Hours) | % of Total Actual Generation |
|---|---|---|---|
| Coal Fired | 337,300 | 1,996 | 48% |
| Natural Gas | 454,611 | 883 | 21% |
| Nuclear | 106, 147 | 806 | 19% |
| Hydroelectric | 77,731 | 255 | 6% |
| Wind | 24,980 | 55 | 1% |
| Ground Solar | 539 | 0.9 | 0.02% |
| Geothermal | 3,251 | N A | N A |
| Other | 100,200 (a) | 162 | 4% |
| Total | 1,104,486 | 4,157 | 100% |

Figure 6.3 A Black Hill near Boulder City, Nevada

184

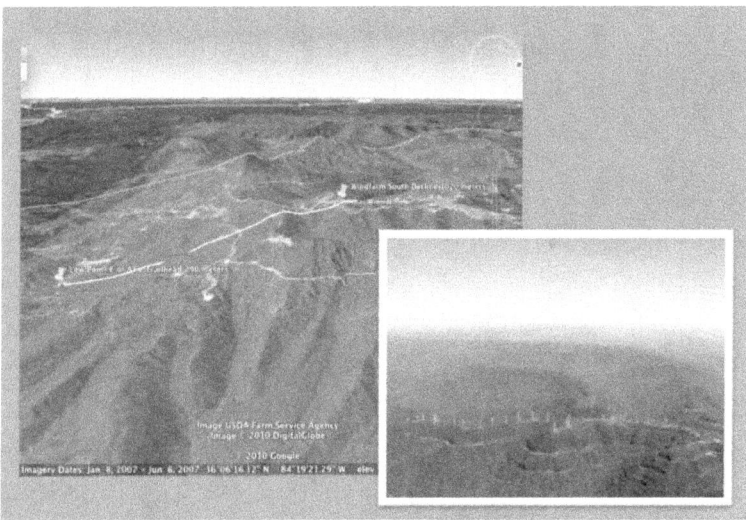

Figure 6.3 B Buffalo Mountain near Oak Ridge, Tennessee

## The Maglev Energy Storage System

The MAPS (Maglev Power Storage) system stores electrical energy by using magnetically levitated and propelled vehicles to move heavy masses uphill. The levitated Maglev vehicles do not physically contact the guideway that they travel along, so that there are no mechanical friction losses. Maximum vehicle speed is moderate, on the order of 100 mph, so that air drag losses are small.

Moving a heavy mass uphill from a lower elevation takes electric power from the grid for the vehicle's magnetic propulsion system. The electrical energy used during the uphill trip is then stored as gravitational potential energy of the mass when it is unloaded at its designated location at the higher elevation. To retrieve the stored energy of the heavy mass, it is then transported downhill by the Maglev vehicles whenever power flow to the grid is desired. During the uphill trip, the vehicle's magnetic propulsion system operates in the motor mode, converting input electrical energy to gravitational potential energy of the mass as it climbs to a

185

higher elevation. During the downhill trip, the vehicle's magnetic propulsion system operates in the generation mode, converting the gravitational energy of the mass back to electric energy which is fed back to the grid. Figure 6.3A & B shows typical locations for Maglev energy storage systems.

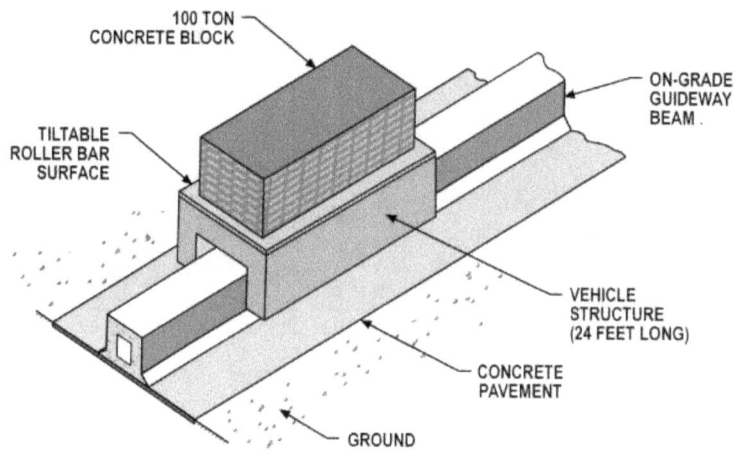

Figure 6.4

Moving a 100 metric tonne mass uphill for 3,000 feet altitude gain stores 250 kilowatt hours of electrical energy. More than 90% of this stored energy can be recovered and fed back to the electrical grid when the block is moved downhill. For easy handling and rapid loading and unloading the heavy mass can be a large concrete block, as shown in Figures 6.4, 6.5, and 6.6. Figure 6.4 shows an isometric view of the Maglev vehicle moving a 100 tonne block along the MAPS guideway. Figure 6.5 shows a cross sectional view of a vehicle carrying the block, while Figure 6.6 shows it after the block has been unloaded.

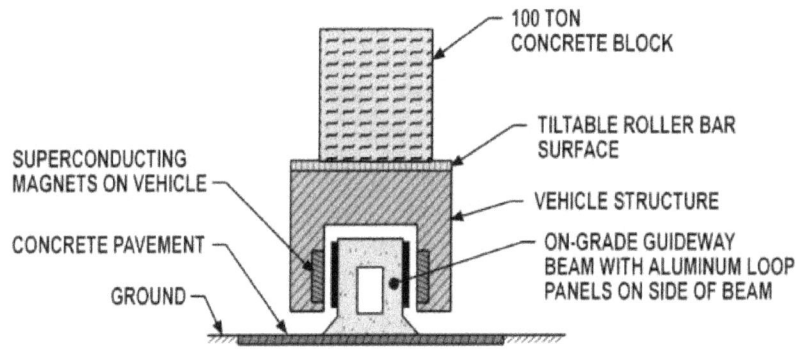

Figure 6.5

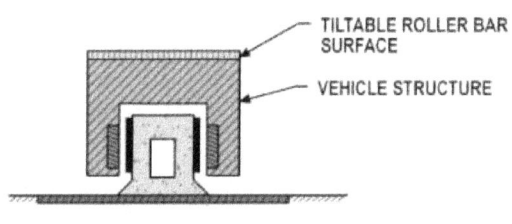

Figure 6.6

The concrete block can be a solid mass of concrete, or a concrete box filled with heavy rocks. The second option would be cheaper, but either is affordable. At 130 dollars per cubic meter for concrete (100$ per cubic yard) and a density of 250 kg per cubic meter, the capital cost of a 100 tonne solid concrete block would be only about 5,000 dollars. Amortized over a 30 year period with 1 storage per day, the cost per kWh stored by the block would be only 1/5th of a cent. The amortized cost of a concrete box filled with rocks would be even smaller, less than 1/10th of a cent per kWh.

The MAPS guideway shown in Figures 6.7 and 6.8 is a simple concrete box beam located on a concrete highway type pavement. The nominal width of the pavement is 20 feet, a bit less than the standard 2 lane highway. The stresses on the pavement when the Maglev vehicle passes a given point are much smaller than highway pavements experience, because the load is spread out over a much larger area. With an 80,000 pound 18 wheeler, the point loads on a highway pavement under the wheels are each on the order of 20,000 lbs per square foot.

The guideway panels contain 3 sets of aluminum loops plus a set of iron plates at the top of the panel. When a Maglev vehicle passes the panel, the iron plates interact with the superconducting magnets on the vehicle, providing a portion of the magnetic lift force. One of the three sets of aluminum coils in the panel carries current induced by the superconducting magnets as they pass the panel, providing additional vertical lift force. The combination of iron plates and set of Figure-of-8 null-flux aluminum loops support the vehicle in the vertical direction. The support is inherently vertically stable – if the vehicle is displaced upwards from its equilibrium suspension height, the magnetic lift force automatically decreases, causing it to move downwards to the equilibrium point. If the vehicle is displaced downwards from its equilibrium point, the magnetic lift force automatically increases, causing it to move upwards to the equilibrium point.

The second set of aluminum loops in the guideway provides automatic horizontal stability. If the vehicle is displaced horizontally in either direction, left or right, from its center position on the guideway beam, force generated by induced currents in the second set of aluminum loops that push the vehicle back to its center position on the beam.

The third set of aluminum loops in the panels carries an applied AC current derived from electric power drawn from the grid. The AC current pushes on the DC superconducting magnets on the vehicle, magnetically propelling it along the guideway. The system can operate either in the motor mode, where the applied AC current propels the MAPS vehicle uphill, or downhill in the generator mode, where the powerful superconducting magnets generate AC power in the guideway loop circuit, which is then fed back to the electric grid.

There is a DC/AC power conditioning link between the MAPS power system and the grid. 60 Hertz power from the grid is rectified to DC and then inverted to AC power at the proper frequency for the MAPS System. As the MAPS vehicle accelerates, the AC frequency is increased – in effect, the MAPS propulsion system operates as a Linear Synchronous Motor. As the MAPS vehicles decelerates, the AC frequency decreases, and the vehicle operates as a Linear Synchronous Generator. In the electric sense, MAPS operates very much like conventional synchronous motors and generators, except that the geometry is linear, not rotary.

The synchronous nature of the MAPS magnetic propulsion system is very important.

The vehicles are phase locked into the AC current wave as it travels along the guideway, with the vehicle speed controlled by the frequency of the AC wave. Variations in external force on the vehicle do not change its speed, only its phase relative to the AC current wave. In effect the MAPS vehicles travel much like a surfer on a water wave, with their speed being the same as that of the wave.

As a consequence, the distance between vehicles operating on a MAPS guideway will automatically remain constant, ensuring that collisions cannot occur. For high propulsion efficiency, the entire MAPS guideway is not completely energized – only those sections on which vehicles are currently operating. The length of each energized block is approximately 200 feet. As a vehicle leaves an energized block, the propulsion power is switched into the next block. Similarly, when the vehicles are running downhill in the generator mode, the output power is switched off from the block that the vehicle is leaving, and switched on from the next block.

Figure 6.9 illustrates the MAPS energy storage mode, in which multiple vehicles can operate on the guideway simultaneously. As each vehicle reaches the top of the uphill guideway, it unloads its 100 tonne block into a storage yard. The flat top of each vehicle has a set of roller bars on which the block moves. Movement can be achieved either by powering the roller bars, with their rotation providing the force required to move the block off of the vehicle, or by hydraulically tilting the upper surface of the vehicle, so that the block slides off onto a conveyer belt, also with roller bars, that moves the block to a designated location in the storage yard. The powered roller bar approach appears to be the most promising system, though the tilt able surface is also practical.

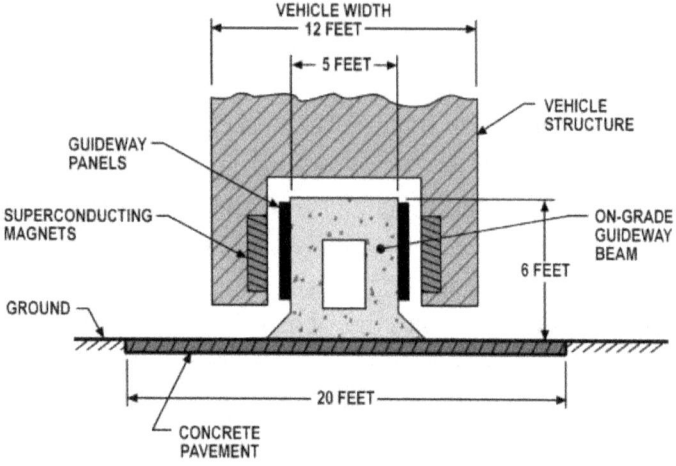

Figure 6.7

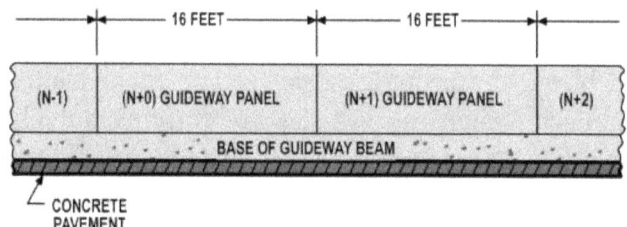

GUIDEWAY PANELS CONSIST OF LOOPS OF ALUMINUM CONDUCTOR ENCASED IN POLYMER CONCRETE

• 1 SET OF ALUMINUM LOOPS PROVIDES VERTICAL LIFT AND STABILITY
• 1 SET OF ALUMINUM LOOPS PROVIDES HORIZONTAL STABILITY
• 1 SET OF ALUMINUM LOOPS PROVIDES MAGNETIC PROPULSION
• PANEL DIMENSION: 4 FEET WIDE, 16 FEET LENGTH, 4 INCHES THICK

Figure 6.8

## ENERGY STORAGE MODE

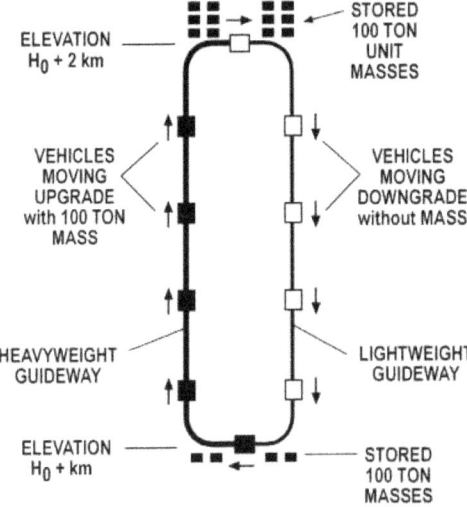

ELEVATION
$H_0 + 2$ km

STORED
100 TON
UNIT
MASSES

VEHICLES
MOVING
UPGRADE
with 100 TON
MASS

VEHICLES
MOVING
DOWNGRADE
without MASS

HEAVYWEIGHT
GUIDEWAY

LIGHTWEIGHT
GUIDEWAY

ELEVATION
$H_0 +$ km

STORED
100 TON
MASSES

Individual
Vehicle Can:

Make 20 Round
Trips/Hour (Site
Dependent)

Store 40 MWH
in 8 Hour Period

Operates @ 30
MW Power
Level

Figure 6.9

A MAPS Maglev vehicle carrying a 100 metric tonne load plus its empty weight of 10 tonnes, for a total weight of 110 tonnes, would have a pavement load of only about 2,000 pounds per square foot, a factor of 10 less than a highway truck. The Maglev vehicle is supported by the magnetic interactions between the super-conducting magnets that are located along the length of the vehicle and the guideway panels on the sides of the box beam

In the power storage mode, the vehicles travel in the clockwise direction on the guideway, while in the energy delivery mode, they travel counterclockwise.(Figure 6.10) The left side of the guideway transports the fully loaded vehicles with their 100 tonne blocks both uphill and downhill, requiring a heavy weight guideway, with the right side of the guideway transports only the unloaded Maglev vehicles which are much lighter than the fully loaded Maglev vehicles, i.e., 10 tonnes versus 110 tonnes.

Multiple vehicles can travel on the MAPS guideway at the same time, either as individual units, or as consists of several vehicles coupled together. A single vehicle can make as many as 20 trips per hour, depending on the height it raises the 100 tonne block, e.g., 3,000 feet, the angle of the guideway, e.g. 30 degrees, and the maximum speed it travels at, e.g. 60 meters per second (134 mph). Vehicles have to decelerate as they approach the upper and lower load/unload points, and accelerate back to maximum speed after they leave them.

## POWER GENERATION MODE

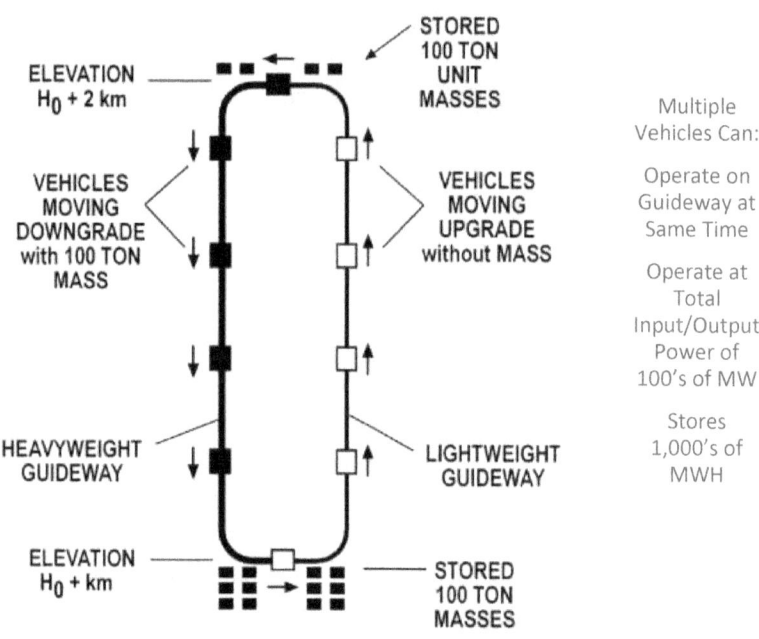

Figure 6.10 MAPS, Power Generation Mode

At a maximum speed of 60 meters per second and a guideway angle of 30 degrees, an individual vehicle carrying a 100 tonne block is operating at 30 megawatts from the grid if it is storing energy and 30 megawatts to the grid if it is delivering energy. The power rating can be reduced if desired, by operating at lower maximum speed.

Similarly, if the guide-way angle is less than 30 degrees, the power rating will be lower.

Over an 8-hour period, an individual vehicle with 20 round trips per hour and an elevation change of 3,000 feet would store 40 MWH. Using multi-vehicle consists, the amount of energy stored would be much greater. For example, operating with a 25 vehicle consist, an energy of 1,000 MWH could be stored/delivered over an 8 hour period, with average input/output of 125 megawatts. Figure 6.11 shows an overall layout of the MAPS facility. There are switching sections at the top and bottom of the guideway to sidings where vehicles currently not in use can be stored. When required to handle increased power levels, either in the storage mode or the delivery mode, vehicles can be rapidly switched out from the sidings to handle the increased power level.

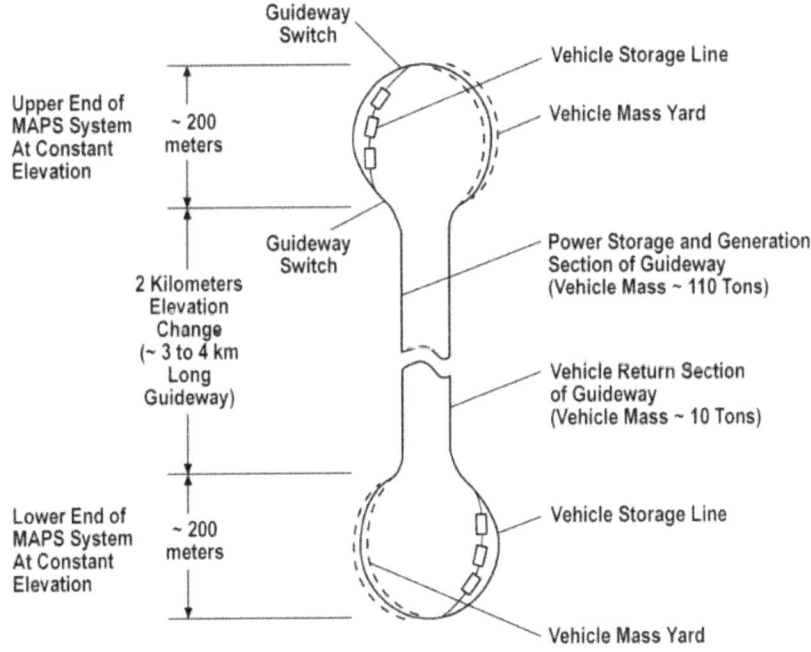

Figure 6.11 Layout of MAPS Energy Storage Facility

MAPS facilities can be sited at a very wide range of locations, in hilly and flat terrains. In hilly terrains, the MAPS guideway would ascend on the rising terrain from a lower elevation to a higher one. On flat terrain, the MAPS guideway could be located in a slant tunnel (Figure 6.12) that descended several thousand feet. Tunnels in hard rock mines descend to a depth of 9,000 feet or more. In coal mines, with less stable rock, tunnels descend to depths as much as 3,000 feet. Alternatively, a vertical shaft guideway could be used, with the blocks stored underground in tunnels, i.e. "drifts" in mining terms, that lead off from the vertical shaft. In certain locations where a greater change in elevation is desired than is available from local hills, the guideway could combine a surface elevation rise with a slant tunnel (Figure 6.12).

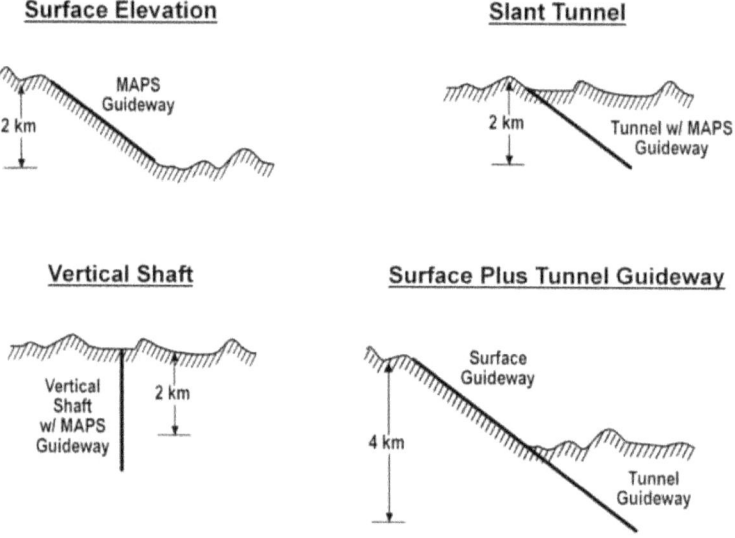

Figure 6.12 Potential Types of Maps Locations

With such a wide range of locations, and an equally wide range of storage capabilities, there is no single design for a MAPS facility. Table 6.1 gives illustrative design parameters for a 1,000 MWH facility with a 3,000-foot elevation change and a somewhat slower maximum vehicle speed of 100 mph than the 134 mph capability. Because of the slower speed, vehicles make only 15 round trips per hour, not 20. MAPS capital and operating costs are derived from detailed design studies carried out for Maglev passenger and freight transport applications. Detailed cost analyses specific to MAPS designs remain to be done.

The largest cost component for MAPS is the cost of operating power for the MAPS facility, which is primarily the cost of propulsion power for the MAPS vehicles as they make their input and output runs. At an overall output/input electrical efficiency of 90%, with grid power purchased at 8 cents/kWh, the power cost component is about 0.8 cents/kWh or $8 per MWH.

Table 6.1
MAPS Storage Capacity and Cost

| Hardware Component | Capital Cost (M$) | Amortized Capital Cost ($/MWH) | Operating Component | Operating cost ($/MWH) |
|---|---|---|---|---|
| Guideway & Storage Yard | 30 | 2.7 | Personnel | 5.4 |
| Vehicles | 80 | 7.3 | Maintenance | 1.0 |
| Power Equipment | 12 | 1.1 | Propulsion | 8.0 |
| Concrete Blocks | 20 | 1.8 | Power | |
| Handling Equipment | 10 | 0.9 | | 1.7 |
| Total | 152 | 12.8 | | 16.1 |
| | | (1.3Cents/kWh) | | (1.6Cents/kWh) |
| Total Cost/MWH= 12.8 + 16.1 = $28.9/MWH = 3 Cents/kWh for Illustrative MAPS System | | | | |

An output/input efficiency of 90% corresponds to purchasing 1.1 kWh of power from the grid for every kWh that the MAPS facility delivers back to the grid. Using a pumped hydro storage unit with a output/input efficiency of 66%, 1.5 kWh of electricity would be purchased from the grid for every kWh delivered back to it. At a cost of 8 cents/kWh purchased from the grid, the cost of the power required to operate the pumped hydro facility would then be 0.5 X8=4 cents per kWh, or $40 per MWH – 5 times greater than the costs of power for MAPS.

The total storage cost for the illustrative MAPS design is then 2.9 cents/kWh ($29/MWH), including the amortized capital costs of the guideway and vehicles, blocks, and the handling equipment, plus the operating costs for power, personnel, and maintenance.

For the range of MAPS applications, depending on site conditions, power storage requirements, etc., the total cost for MAPS storage will be in the range of 2 to 3 cents/kWh ($20 to $30 per MWH). This storage cost will be well below the cost of power generation from wind and ground solar sources. Accordingly, MAPS can be used for storing energy from renewable wind and ground solar plants.

Finally, the presently estimated cost of the MAPS guideway is based on trucking prefabricated guideway beams to the construction site and placing them on a pre-poured concrete pavement, similar to the construction mode for the Maglev elevated monorail guideway for transport of passengers and freight. The guideway cost could significant-ly reduced by casting the concrete guideway box beam directly in place on the concrete pavement.

## Status of Maglev Energy Storage Technology

MAPS uses superconducting Maglev technology, invented by Powell and Danby in 1966. (5) Based on their inventions, Japan has demonstrated that superconducting Maglev is a practical mode of transportation. Japan Railways 1st generation passenger transport system (Figure 6.13) is now operating in Yamanashi, Japan. The 27 mile long 2-way Maglev demonstration route has carried 100,000 passengers safely and reliably, with accumulated running distances of hundreds of thousands of miles. Japan Railways plans to incorporate the Yamanashi line into a 300 mile Maglev route between Tokyo and Osaka that will carry 100,000 passengers daily, with a trip time of 1 hour. The Japanese superconducting Maglev system holds the World speed record for ground transport, at 370 mph.

The superconducting Maglev System is inherently strongly stable, with magnetic restoring forces that automatically oppose any external force (e.g., winds, curves, etc.) that would act to displace the vehicle from its equilibrium suspension position. The gap between vehicle and guideway for *superconducting Maglev* is typically about 10 cm. The large gap enables the guideway to be built with large tolerances, reducing the cost of construction.

The electromagnetic maglev is not practical for MAPS, because the required electric power to operate its electromagnets would be far too great if it were used to transport 100 tonne blocks uphill and downhill for MAPS energy storage. The superconducting magnets used on superconducting Maglev vehicles have zero electrical energy losses, and lift much heavier loads than conventional electromagnets. There is a small energy requirement for the refrigeration equipment of the cryogenic superconducting magnets, but this power requirement is tiny compared to the input/output power that the MAPS vehicles handle to move blocks uphill and downhill.

Figure 6.13 View of Japan Railways Superconducting Maglev Guideway and Vehicle System

Figure 6.14 Artist's Drawing of Maglev 2000 Vehicle on Monorail Guideway

Powell and Danby's new 2nd generation Superconducting Maglev 2000 system (Chapter 4) is designed to have a low guideway construction cost, about 30 million dollars per mile, a factor of 2 or more lower than 1st generation systems. To achieve this goal, low cost prefabricated monorails are used for most of the elevated guideway construction. Figure 6.14 shows an artist's view of a Maglev 2000 passenger vehicle on the elevated monorail guideway.

The prefabricated monorail beams would be mass-produced in factories, with their guideway loop panels, sensors, electronic equipment, etc. attached to them at the factory. The beams and piers would then be transported by truck or rail to the construction site, where they would be quickly erected on pre-poured concrete footings or pilings, using conventional cranes. Guideway cost would be kept low by the use of conventional box beams for the monorail, which minimizes the amount of materials required, and prefabrication, which minimizes expensive field construction.

The ability to operate on a planar guideway as well as monorail also helps to reduce the construction cost of Maglev routes. When operating in densely populated urban and suburban areas,, the Maglev 2000 vehicles does not need to build a new, very expensive guideway with its accompanying disruptions and modifications to existing infrastructure. Instead, Maglev 2000 can transition to, and operate on, existing RR tracks to which aluminum loop guideway panels have been attached on the cross-ties. Conventional trains can continue to use the RR tracks, given appropriate scheduling. The cost of attaching guideway panels to enable levitated Maglev 2000 operation is very small, only about 8 million dollars per mile, compared to the high cost of a new elevated guideway for the 1st generation Maglev systems that cannot operate on existing RR tracks.

The Maglev 2000 vehicle has superconducting quadrupole magnets that induce currents in aluminum loop panels on the guideway beam. The magnetic interaction between the vehicle's superconducting magnets and the induced currents in the aluminum guideway loops stabilize and levitate the vehicle. The magnetic levitation force automaticly opposes any external force (wind, curves, etc., that try to displace the vehicle from its equilibrium levitation position. An AC current in one set of the aluminum guideway loops magnetically propels the Maglev 2000 vehicle.

Figure 6.15 shows one of the two wound superconducting loops used for the Maglev 2000 quadrupole. The loop has 600 turns of NbTi superconducting wire, supplied by SuperCon, Inc. of Shrewsbury, Massachusetts. At the design current of 1,000 Amps in the NbTi wire, the Maglev 2000 Quadrupole has a total of 600,000 Amp turns in each of its 2 superconducting (SC) loops. The SC winding is porous, with small gaps between the NbTi wires to allow liquid Helium flow to maintain their temperature at 4.2K, and to stabilize them against flux jumps in micro-movements.

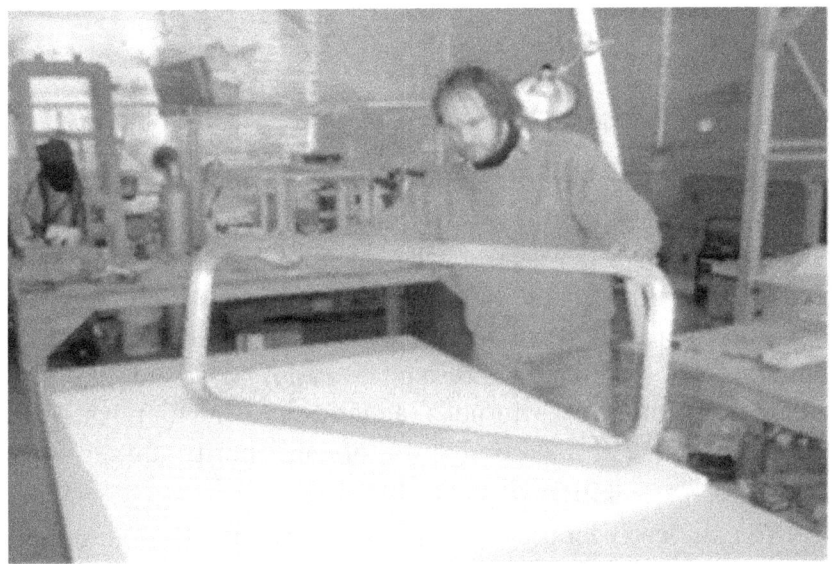

Figure 6.15 NbTi Superconductor Loop for Maglev 2000 Quadrupole

Figure 6.16 Shows the SC loop enclosed in its stainless steel jacket

Figure 6.16 shows the SC loop enclosed in its stainless steel jacket. Liquid Helium flows into the jacket at one end and exits at the end diagonally across from the entrance providing continuous Helium flow through the SC winding. Before insertion of the SC loop into the jacket, it is wrapped with a thin sheet of high purity, aluminum (5,000 residual resistance ratio) to shield the NbTi superconductor from external magnetic field fluctuations. After closing the jacket, a second layer of high purity aluminum is wrapped around it for additional shielding.

The guideway loop panels contain 3 sets of wound aluminum loops, composed of a set of four Figure-of-8 loops, a set of 4 dipole loops, and 1 long LSM propulsion loop. The aluminum conductor has a ~10 mil layer of nylon using a dip process to coat the conductor. The nylon insulation withstood 10 Kilovolt tests without breakdown. Figure 6.17 shows a completed guideway loop panel with all of its loops.

Figure 6.17 Completed Guideway Panel with Figure-of-8, Dipole, and LSM Propulsion Loops

The completed panel is then enclosed in a polymer-concrete structure for handling and weather protection. Polymer concrete – a mixture of aggregate, cement and plastic monomer – can be cast into virtually any form as a slurry. When the monomer polymerizes (the rate of polymerization is controlled by the amount of added promoter), the resulting concrete-like structure is much stronger – a factor of 4 or greater – than ordinary concrete and not affected by freeze-thaw cycles, salt, etc. Figure 6.18 shows a completed partial section of a polymer concrete panel left outside of the Long Island facility for 2 years. It was subjected to a wide range of weather conditions and multiple freeze-thaw cycles over the 2 year period, without any degradation

After being fabricated at the Maglev factory, the guideway panels would be attached to the side of the MAPS beam to be shipped to a MAPS construction site. Alternatively, the guideway box beam could be poured in place at the site, and the guideway panels, which would be trucked to the site from the factory, attached to the in-place beam.

Figure 6.18 Polymer Concrete Panel with Enclosed Aluminum Loop

Figure 6.19 Photo of 72 Foot Long Monorail Guideway Beam Delivered to Maglev 2000 Facility in Florida from Construction Site in New Jersey

For the elevated guideway transport system, the monorail beam is a hollow box beam made with reinforced concrete. Beam length is 22 meters and weight is 34,000 kg. It uses post tension construction, which allows the tensioning cables in the base of the beam to be re-tightened if some stretching were to occur. The beam is tensioned to have a 0.5 cm upwards camber at the midpoint of the beam when it is not carrying a Maglev vehicle. When the Maglev vehicle is on the beam, the beam flattens out to a straight line condition, with no vertical dip or camber along its length. Then placed on-grade for the MAPS system, the guideway beam is fully supported along its entire length, not just at its ends, the case for the elevated guideway. As a result the MAPS guideway beam will be simpler and less expensive than the elevated beam.

Figure 6.19 shows a photo of the fabricated elevated guideway beam after transport by highway truck from the manufacturing site in New Jersey to Maglev 2000's facility in Florida. No problems in transport by highway were encountered.

Fabrication and testing of the basic Maglev 2000 components – superconducting quadrupole magnets, aluminum loop guideway panels, monorail guideway beam, and vehicle body – have been successfully carried out. The next step for the development of the commercial 2<sup>nd</sup> generation Maglev 2000 system is to test operating vehicles on a guideway.

The MAPS system uses the same components – Superconducting quadrupoles, monorail guideway beam (on-grade, rather than elevated, however), aluminum guideway loop panels for vertical lift, horizontal & vertical stability, and magnetic propulsion, and a planar surface electronic switch between the main guideway and sidings.

The principal difference between the Maglev 2000 transport guideway and the MAPS guideway is the addition of iron plates in the guideway panels. The iron plates provide a substantial fraction of the lift force, without having to induce currents in the aluminum guideway loops. This reduces $I^2R$ losses in the aluminum loops, increasing propulsion efficiency. The MAPS suspension is still inherently stable, because the stability forces from the null-flux aluminum loop circuits are greater than the destabilizing forces associated with the iron plates. (6)

The MAPS vehicles are much simpler and cheaper than these used by the Maglev 2000 System to transport passengers, autos, highway trucks, and freight containers. They are simple sled type structures with attached superconducting magnets and a compact cryogenic refrigeration unit and its associated plumbing. The powered roller bars on the upper surface of the sled are similar to units that operate in many industrial facilities. Overall, MAPS would use virtually the same technology as Maglev 2000's transport system except that it would be simpler and lower in cost.

Replacing expensive electricity from natural gas peaking plants with much less expensive power from baseload plants that used MAPS would be a very large market. In the US, storing 883 million MWH (the MWHs generated by natural gas power plants) at an average cost of 3 cents/kWh would bring in 28 Billion dollars annually to MAPS, sufficient to recover costs and pay back the cost of the MAPS facilities over a 30 year period.

Assuming that the cost of peaking power from natural gas plants was 9 cents/kWh greater than the cost of baseload power, that would then correspond to a production power savings of 6 cents per kWh using MAPS (+9-3 = 6 cents saved), a net savings of 56 Billion dollars annually. If the cost differential of peaking power from natural gas, compared to baseload power, was greater than 9 cents/kWh, the net savings using MAPS would be even greater than 56 Billion dollars per year.

To replace with wind and ground solar all US power now generated from coal and natural gas power plants, which account for virtually all of US power generated using fossil fuels and store the output power from wind and ground solar using MAPS, would cost 88 Billion dollars for MAPS storage, based on 3 cents per kWh. The corresponding cost of $CO_2$ emissions that were eliminated would be about 40 dollars per ton – a reasonable price to pay.

The corresponding total 2008 World market for MAPS would be about 5 times greater than in the US, since the total World generation is approximately 5 times greater than the US market. The 2008 MAPS revenue for eliminating World natural gas power generation would be 130 Billion dollars per year; for eliminating coal and natural gas, the revenues would be 370 Billion dollars per year.

The EIA (Energy Information Administration) projects that World electrical generation will double by 2035 (7) as World population increases and countries become more industrialized. Coal and natural gas power plants will continue to dominate electric production, accounting for almost 70% of total power production.

It is disheartening that wind and ground solar power play such a minor role in EIA projections for 2035, only about 2% of the total. To achieve the goal of major reduction in World $CO_2$ emissions, wind and ground solar will have to play a much greater role. They can do so, using MAPS storage to enable wind and ground solar to reliably deliver large amounts of power on demand.

The annual revenues shown for MAPS are very large, and correspond to a major World industry with many economic and environmental benefits.

## Summary and Conclusions

MAPS (MAglev Power System) is a new way to store large amounts of electrical energy at very low cost and very high electrical efficiency that can make wind and ground solar become the major energy supplier. MAPS can store thousands of Megawatt Hours at very low cost, 2 to 3 cents per kWh stored, and has an output/input electrical efficiency of over 90% -- that is, for 100 kilowatt hours (kWh) of electrical energy fed into a MAPS storage facility, over 90 kWh is returned to the grid on demand. MAPS stores electrical energy from the grid by using levitated and magnetically propelled Maglev vehicles to move heavy masses from a lower elevation to a higher elevation. The input electric energy to the Maglev propulsion system, which operates in the motor mode as it moves mass uphill, is stored as gravitation potential energy of the mass as it rests at its elevated location.

To convert the stored gravitational energy of the mass back to electric energy and return it to the grid, the Maglev vehicles moves the mass back to the lower elevation, with the Maglev propulsion system operating in the generator mode.

There are no mechanical friction losses in moving the storage masses uphill and downhill, the air drag losses are small, and the Maglev propulsion system is very efficient, resulting in a high electrical efficiency of 90% or better, output electrical energy deliver back to the gird, divided by input electric energy from the grid.

MAPS uses Maglev technology that has been developed as a practical, very reliable system for high speed transport of passengers. No breakthroughs are required – only adapting the basic existing Maglev technology for a new application. MAPS can be demonstrated and certified for commercial use relatively quickly.

MAPS has important advantages compared to other electric energy storage systems. Pumped hydro, the most widely used energy storage system has a much lower output/input electrical efficiency. Moreover, pumped hydro has environmental restrictions and problems, and is very limited where it can be sited. In contrast, MAPS has far fewer environmental problems, and can be sited at a much wider range of locations. Similarly, MAPS is more efficient than the compressed air storage systems, and can be sited at a much wider range of locations.

Other energy storage systems, i.e., batteries, flywheels, superconducting magnetic energy Storage (SMES) are much more expensive per kWh stored than MAPS, and are only suited for storing relatively small amounts of electrical energy for dynamic stabilization of the electrical grid in the event of frequency and for mismatches in phase. They are not practical for storage of large blocks of electric power.

# Chapter 7

# Clean Synthetic, Non-fossil Fuels from Air & Water

*"Whither goest thou, America, in thy shiny car in the night."*
Jack Kerouac

Questions to the reader. What do you and your automobile have in common? Flesh vs metal and plastic? No. Wheels vs legs? 100 mph Speed vs 4 mph? No. Intelligence and the ability to communicate? No, though Google's self-driving autos are on the way there.

Both us and cars have one thing in common. We both take in fuel for energy and excrete waste products when we burn the fuel. Sure, the fuel is different – beef and chicken, milk and donuts, french fries, etc., for humans, vs gasoline or diesel for cars – but leaving aside certain human waste products, both cars and humans expel carbon dioxide and water vapor generated by burning fuel or food.

Water vapor is harmless. But carbon dioxide can be dangerous. Earth's temperature depends on its concentration in the atmosphere, as described in chapter 2.

How much carbon dioxide ($CO_2$) do humans and cars emit? On average, each human breathes out 2.3 pounds (1 kilogram) of $CO_2$ per day (1), equal to 0.365 metric tonnes, annually, from the digestion of the food they eat. A metric tonne is 1,000 kilograms. Cars emit much more. Burning 1 gallon of gasoline emits 19.65 pounds (8.9 kilograms) of $CO_2$: burning 1 gallon of diesel fuel emits 22.3 pounds (10.1 kilograms) of $CO_2$(2).

The average American drives his/her car about 10,000 miles per year, averaging 27 miles per day. Assuming that it is a fuel efficient car 25 mpg, and not some gas guzzler, that's about 1.1

gallons of gas per day. The car's CO2 emission is approximately 10 kilograms daily, 10 times more than its driver.

Not only is the car's CO2 emission much greater than the driver's, there is an extremely important difference in how the two kinds of CO2 emission affect the Earth's environment.

Human CO2 emissions into the atmosphere are taken up by the plants that make the food that humans eat. There is no net addition of CO2 into the atmosphere. The environment is in balance – animals, including humans, eat plants, releasing CO2, which is taken up by the plants that grow the food for the animals. It's sustainable forever, as long as something does not disturb it.

Well, the CO2 emitted by cars does not come from today's Nature, but from immense amounts of fossil fuels deposited in the Earth millions of years ago. And it's being consumed at rates that are too rapid for today's Nature to handle.

As a result, since the beginning of the Industrial Revolution, atmospheric CO2 concentration has increased from 280 ppm (parts per million) to today's 400 ppm (1), and is increasing by approximately 2 parts per million per year. By 2100, with an increase in World population from today's 7 Billion people to 9 Billion, and the large increase in economic activity of much of the population, CO2 atmospheric concentration is likely to be on the order of 600 ppm or more. As described in Chapter 3, if we keep burning fossil fuels that would be an environmental catastrophe. Current World CO2 emissions from fossil fuels are about 32 Billion metric tonnes per year (3).

Transportation is fundamental to modern society. We must have cars, trucks, airplanes, trains, ships, pipelines, tractors, etc., to bring us food and goods, let us travel for work and pleasure, and militarily defend ourselves. Without modern transport we would be back to horses, oxen drawn wagons, sailing ships, and walking – life as it was a few hundred years ago, as illustrated in Chapter 1.

And today's transport consumes a great deal of fossil fuel. Figure 7.1 (4) shows United States CO2 emissions from fossil fuels by Energy sector in 2011. Electric power was the biggest (40%), followed closely by transport (34%), with the other sectors residential, commercial, and industry totaling only 27%. Most of the US transport $CO_2$ emissions come from cars and trucks. Of the total 1.84 Billion tonnes of $CO_2$ from transport, 1.095 Billion tonnes came from gasoline and 0.427 Billion tonnes from diesel fuel. The remaining 0.3 Billion tonnes came from airplanes, ships, trains, etc. It is not practical to capture and sequester the $CO_2$ emissions from the multitude of cars, trucks, airplanes, ships and trains. They are inevitably released to the atmosphere.

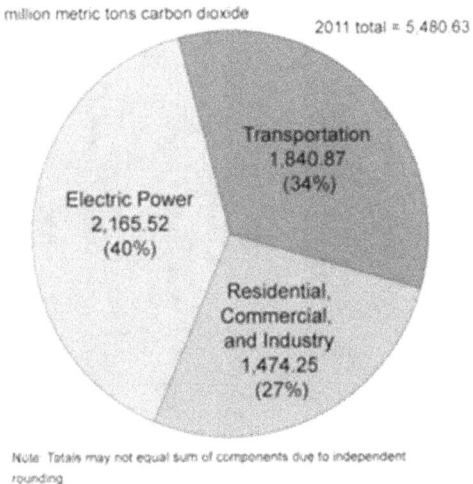

**U.S. Energy-Related Carbon Dioxide Emissions by Sector, 2011**

million metric tons carbon dioxide

2011 total = 5,480.63

Transportation 1,840.87 (34%)

Electric Power 2,165.52 (40%)

Residential, Commercial, and Industry 1,474.25 (27%)

Note: Totals may not equal sum of components due to independent rounding

Source: U.S. Energy Information Administration. *Monthly Energy Review.* Tables 12 2-12 5 (May 2012)

Figure 7.1

Petroleum is by far the biggest fuel for transport. Figure 7.2 (5) shows where energy for the US economy comes from and where it goes. Petroleum is the biggest source of energy, 37% of total energy input. Natural gas trails at 25%, followed by coal at 21%, with renewables (wind, solar, hydropower and biomass) far behind at 8%, essentially tied with nuclear (9%). 72% of petroleum energy goes to transport, with it constituting 94% of the energy consumed by transport. Renewables, i.e., ethanol and a tiny amount of biodiesel, account for only 3% of transport energy input, tied with natural gas, also at 3%.

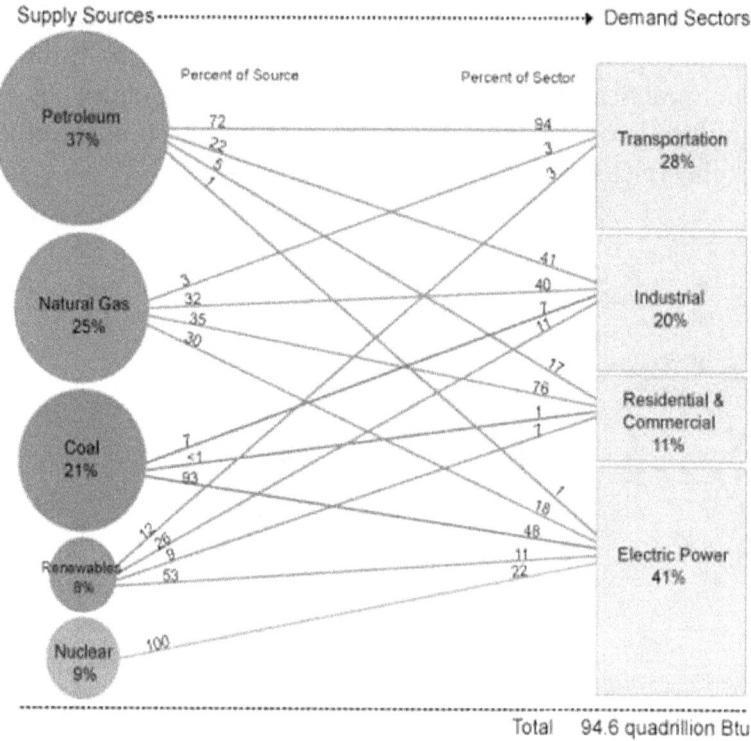

Figure 7.2
US Energy Flows: 2009

The global transport picture is pretty much the same as the US, though less intense. The US with 310 million people consumes 18.5 million barrels of oil daily (6), 72% of which goes for transport. The other 6,700 million people in the World consume 73.5 million barrels per day for a World total of 92 million barrels per person. (6) 61.5% of which goes for transportation (7).

So the US uses 13.3 million barrels of oil per day for transport. The rest of the World uses 45.2 million barrels per day for transport. In per capita terms, the average American uses 1.8 gallons per day of oil for transport. The average for the rest of the World is 0.28 gallons per capita per day.

Conclusion? The ratio of US oil consumption per capita for transport to the rest of the World's per capita consumption is more than 6/1. Further conclusion? As the less developed countries like China, India, and others continue on their paths to bigger GDPs and greater industrialization, their transport needs and energy consumption will grow.

There are currently approximately 1 Billion cars in the World. By 2050 AD, it is projected that there will be 2.5 Billion automobiles, with a World GDP almost a factor of 3 greater than today's 70 Billion dollars GDP. Annual passenger miles and freight ton miles will be double today's values.

Figure 7.3 shows the number of automobiles per 1,000 population in a country as a function of the GDP per capita for the country. The relationship is essentially linear. At $50,000 GDP per capita, there are approximately 800 automobiles per 1,000 persons. At $10,000 GDP per capita, 5 times lower, 160 automobiles per 1,000 population. 800 cars per 1,000 population approaches the limit, although some Americans do own 2 or more personal cars.

As World GDP grows, so will the number of World cars. Average World GDP is projected to grow from about 70 Trillion dollars today to over 180 Billion dollars in 2050 AD. (8) Combined with the population increase from 7 Billion today to 9 Billion in 2050 AD, the per capita World average GDP will grow from $10,000 today to $20,000 in 2050, a factor of 2 increase. The average number of cars per 1,000 population also increases by a factor of 2 from 140 to 280.

Combining today's CO2 transport emissions from the US, 1.8 Billion tonnes per year, and 7.2 Billion tonnes from the rest of the World, we get a total of 9 Billion tonnes per year from transport, 28% of total global CO2 emissions – almost 1/3 rd.

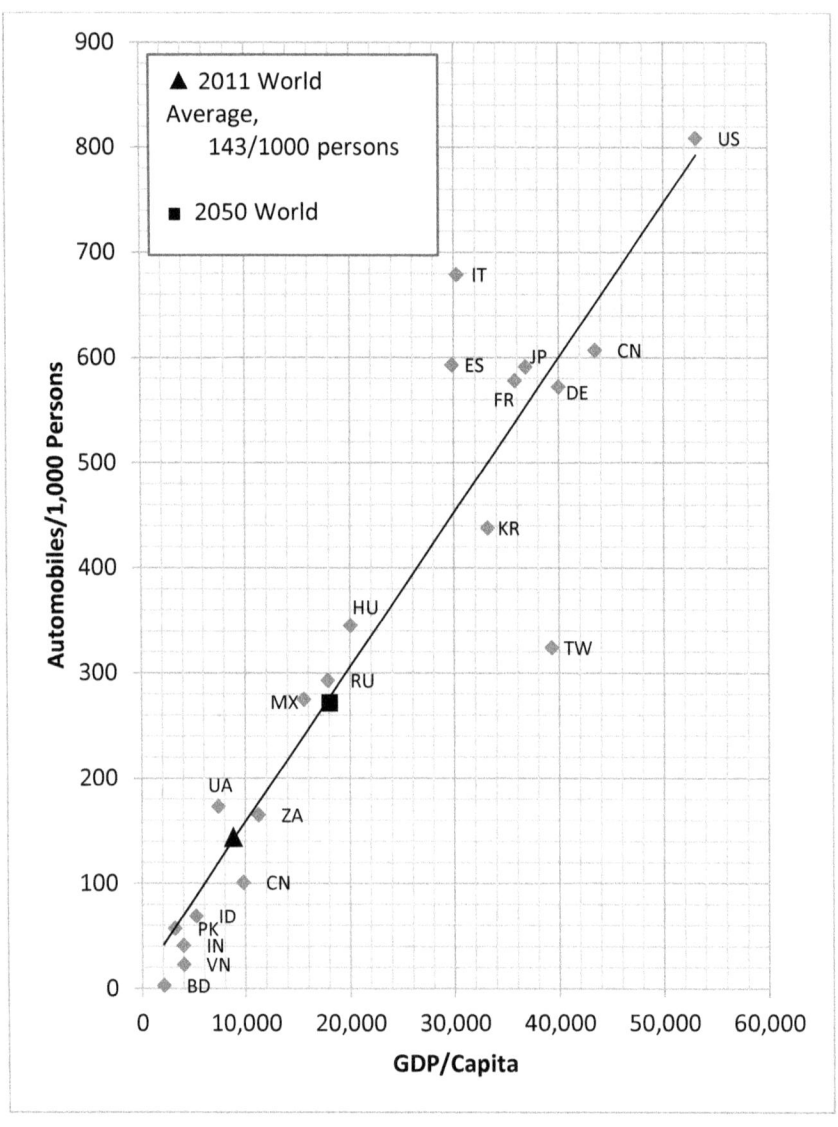

Basis: 2010 – 2013 Data PPP (Adjusted)

Figure 7.3
Automobiles Per 1,000 Persons as a Function of GDP Per Capita for Different
Countries

If transport doubles over the next 35 years to 2050 AD, World transport $CO_2$ emissions will also double, to about 18 Billion tonnes per year, if transport continues to depend on oil fuel. We could not maintain atmospheric $CO_2$ concentration at 400 ppm, to avoid environmental catastrophe.

If we are, to maintain, and hopefully grow, World living standards, we must continue and grow our transport systems. Without them, modern society would collapse. On the other hand, if we continue to use fossil fuels for transport, we will greatly increase atmospheric $CO_2$ concentration, causing environmental catastrophe and the collapse of modern society.

Conclusions. Humanity has a real choice to make now. If we want to avoid environmental catastrophic and social collapse, we must begin to transition to non-fossil fuel transport as soon as possible. We must acknowledge reality, not deny it. It is ironic. If some species were emitting 32 Billion tonnes of $CO_2$ per year, increasing its atmospheric concentration to dangerous levels, humanity would take action, hunting down and destroying the species that was emitting the $CO_2$. But since humans are doing it, we deny it and take no real action. That's human nature.

What are the alternatives to fossil fuels for transport? Are they feasible? Practical? Too expensive? Limited capability? Limited applicability? Environmentally harmful? The alternatives to fossil fuels for transport are:

- Biofuels
- Hydrogen fuel
- Batteries, i.e. "electric" cars
- Electrically powered Maglev
- Synfuels from air and water

Biofuels are feasible, but not practical. Growing Biofuels competes with growing food. Already, on the order of 1 Billion people in the Word are hungry and malnourished. Total US consumption of liquid fuel, including gasoline, diesel fuel, and jet fuel is about 200 Billion gallons annually. To produce the equivalent amount of liquid fuel with 1.5 gallon of ethanol equivalent to 1 gallon of gasoline, leaving aside the extra energy required to produce it, would take 770 million acres of US farmland, 3 times greater than the 300 million acres we grow our food on – completely impossible. Conclusion? It is silly to expect that biofuel will be able to meet more than a very tiny fraction of future World transport needs.

Hydrogen fueled cars also are not practical. They sound great – clean energy, only water out of the tailpipes, etc. – but the realities do not support hydrogen fuel in cars. Yes, hydrogen can be produced by electrolyzing water using low cost electricity, and we propose that for synthesizing gasoline, diesel fuel and jet fuel from air and water.

But, using hydrogen as a fuel stored in automobiles and trucks is extremely frightening. Imagine 230 million cars and trucks driving 70 mph on US highways, each with an insulated tank of very cold liquid hydrogen, or a tank of compressed hydrogen gas at 5,000 psi. If the car is seriously damaged in an accident, and there are many thousands of such accidents every year on US highways. Hydrogen gas could leak from the tank, mix with air, and explode, with a force of hundreds of pounds TNT causing other nearby cars to explode.

Or, when filling your car with hydrogen at a filling station, and there is a hydrogen leak, from your car, another car, the pump, or the station storage tank that causes a hydrogen explosion. It doesn't matter who is responsible – you're dead. Or, you park your hydrogen car in an underground garage along with dozens of other hydrogen cars. One car leaks and explodes, causing all the other cars to explode and bringing down the building with its occupants.

Hydrogen fueling the World's 2.5 billion cars in 2050 AD? No way.

In contrast, electric cars are practical. In fact, electric cars have been driving around American for well over 100 years. Figure 7.4 shows the Parker electric car, produced from 1899 to 1915. It had a top speed of 14 mph and a range of 50 miles (10). Figure 7.5 shows Thomas Edison with an electric car in 1913.

Electric cars were popular at the beginning of the 1900's. 40 percent of US automobiles were powered by steam, 38 percent by electricity, and 22 percent by gasoline. There were 33,842 electric cars registered in the United States (10). There even was a battery swapping service provided by a subsidiary of General Electric. Well-to-do upper class families loved their electric cars. They were comfortable, luxurious, noiseless, didn't smell like gasoline cars, didn't require long start-up times – 45 minutes in cold morning for Stanley Steamers – and didn't require a hand crank for startup, as did gasoline cars.

James Jordan's father started his working life as an 11 year old chauffeur in a Baker electric car driving an upper class family around in Winston Salem, North Carolina. Electric cars were preferred over gasoline cars in the town. Everybody loved electric cars.

However, things started to change, and gasoline cars took over. Charles Kettering invented the electric starter in 1912 (10) which eliminated the hand crank, a really big deal, and gasoline cars started to use mufflers. Gasoline became much more plentiful, and Henry Ford invented the assembly line, which made gasoline cars much cheaper. By 1912 an electric car cost twice as much as a gasoline car. (10)

The electric car industry withered and died. From time to time, there were efforts to revive it, but they did not succeed. Figure 7.6 shows the 1960 Henny Kilowatt, based on the European Renault Dauphine. Top speed was 60 mph, with the capability to travel for almost an hour on a single charge (10). A good car, but it did not last long. Production stopped in 1961.

Electric cars began their revival in the 1990's. Today, there are lots of makes and models, both all-electric and electric/gasoline hybrids. Manufacturers include BMW, Mercedes Benz, Chevrolet, Honda, Nissan, Fiat, Mitsubishi, Kia, Ford, Toyota, and Tesla.

Tesla electric cars receive the most media attention and are top rated. Figure 7.7 shows a photo of the Tesla roadster. However, the other electric car makes and models are also excellent automobiles.

Figure 7.4 Thomas Parker Electric Car

Figure 7.5 Photo of Thomas Edison with Electric Car 1913

Figure 7.6 1960 Henney Kilowatt Electric Car

Figure 7.7 Tesla Roadster

Electric vehicles still constitute a very small percentage of new car sales Worldwide. The top 10 countries that had the largest market share of all-electric cars in 2013 are listed below, along with the percentage of new car sales that were all electric vehicles. (11)

| Rank | Country | All-Electric Car Market Share (2013) |
|---|---|---|
| 1 | Norway | 5.75% |
| 2 | Netherlands | 0.83% |
| 3 | France | 0.79% |
| 4 | Estonia | 0.73% |
| 5 | Iceland | 0.69% |
| 6 | Japan | 0.51% |
| 7 | Switzerland | 0.39% |
| 8 | Sweden | 0.30% |
| 9 | Denmark | 0.28% |
| 10 | United States | 0.28% |

All of the above 10 countries have high per capita incomes, yet electric car sales are very small – growing, yes, but not rapidly.

In the US, 250,000 plug-in electric cars have been sold since 2008, about 40% of them all-electric cars and 60% plug-in hybrids. That's equivalent to 1/1000th of the approximately 250 million cars on US highways. (11)

Worldwide, about 300,000 all-electric cars and light utility vehicles have been sold since 2008. That is 1/3000th of the Billion cars now operating Worldwide.

This very small fraction of the car market for electric cars, and the very small fraction of the World's total number of cars, raises the question,

Can electric cars become a major part of long-term, sustainable transport for the World?

While electric cars will increase as a fraction of the World total cars in 2050, it is very unlikely that a majority of the projected 2.5 billion cars will be electric. Many drivers will still prefer gasoline or diesel powered cars, because of their drawbacks.

- Electric cars have much shorter range than gas powered cars – typically 100 miles vs 400 miles.
- It takes many minutes to recharge an electric car. To refill a gas tank takes only 1 minute.
- There are 250,000 gas stations in the 48 states. Electric charging stations? Much fewer. Searching for a charging station if your car battery is almost dead? Good luck.
- Electric cars cost more than gasoline cars. Battery cost is on the order of $100 per mile of driving range.
- Cooling the inside of an electric car when its hot outside takes a lot of power, reducing driving range, $10,000 for 100 miles.
- Batteries catch fire sometimes, especially lithium ion batteries, the favored battery for electric cars.
- Batteries are expensive, and can fail after long-term use, both in years and miles driven. Replacing them will be costly.
- There are resource limits on the material used to manufacture batteries, particularly lithium. Enough for 2.5 Billion autos? Not certain, since battery lithium would have to be recycled many times for sustainable long term transport, with losses for each recycle. Also, there are important other uses for lithium.

Thus, it appears very likely that many people will choose gas ,and diesel powered cars over electric cars. And they will want to have gas and diesel fuel even if it comes from fossil sources. In addition, hydrocarbon fuel will still be needed for airplanes, commercial jet liners, and military aircraft – electric airplanes are not likely—trucks, diesel powered locomotives, etc. So we will still need liquid hydrocarbon fuels for the long-term.

Currently, we consume 21 Billion barrels of oil per year for transport, emitting 9 Billion tonnes of carbon dioxide annually from cars, trucks, airplanes, ships and locomotives. Transport volume will double by 2050. If we continue transport's dependence on fossil fuels, carbon dioxide from emissions from transport would increase to 18 billion tonnes per year – one half of today's total CO2 emissions.

Maglev and electric cars will substantially reduce our need for oil for transport but won't eliminate it. There still will be demand for liquid hydrocarbon fuels for transport.

Of the 33 billion barrels consumed annually be the World, 21 Billion go for transport. If we stick with oil we will need 40 billion barrels per year for transport in 2050 AD.

With Maglev and electric cars we will still require on the order of 10 billion barrels per year of liquid hydrocarbon fuel, 1/4th of transport demand in 2050.

## Synthetic Fuels from Air and Water

Is there an alternative to obtaining these liquid hydrocarbon fuels from fossil fuels, either from crude oil or by synthesis from natural gas or coal? Yes, there is – by reacting carbon dioxide extracted from the atmosphere with hydrogen obtained by electrolyzing water. Consumption of the resulting synthetic fuels will not cause an increase in carbon dioxide concentration in the atmosphere, since CO2 from burning the synthetic fuel is offset by CO2 extraction from the atmosphere to make the synthetic fuel.

We now turn to describing how synthetic gasoline and diesel fuel can be manufactured at prices competitive with present fuels derived from crude oil. Carbon dioxide extracted from the Earth's atmosphere, plus hydrogen from the electrolysis of water, with low-cost electric power from space solar power satellites launched by StarTram, as described in Chapter 5, is used.

There is nothing new about making synthetic fuel, both gasoline and diesel, without using oil. The Fischer-Tropsch process, invented in 1925 (12), reacts carbon monoxide (CO) and hydrogen (H2) to make a range of hydrocarbons that can be processed into gasoline and diesel fuel. Alkanes, the straight line hydrocarbons, are produced by the chemical reaction shown below.

$$(2n+1) \ H_2 + nCO \rightarrow C_n(H_2)_{n+1} + nH_2O$$

Octane, a major component of gasoline, has 8 carbon atoms, with the formula C8H18. Diesel fuel contains higher molecular weight hydrocarbons, e.g. decane, C10H22 and above. Gasoline and diesel fuel are complex mixtures of many different hydrocarbons, both straight line molecules, branched molecules, and aromatic molecules. Process parameters and product refining are controlled so as to obtain the desired type of synthetic fuel.

Because of fuel shortages for its army and air force in World War II, Nazi Germany produced large amount of synthetic fuels using the Fischer-Tropsch process. In early 1944 German synthetic fuel production reached more than 124,000 barrels per day. (13)

Figure 7.8 NREL FT diesel vs conventional diesel photo

Worldwide, currently 260,000 barrels per day of synthetic liquid fuels are produced using the Fischer-Tropsch process and other processes that react syngas, i.e. a mixture of carbon monoxide (CO) and hydrogen (H2) to manufacture the synthetic fuel 260,000 barrels per day. (13) That's 100 million barrels per year, 1% of our goal of 10 Billion barrels per year of manufacturing synthetic gasoline and diesel fuel from air and water. It shows that the process is practical and affordable if the cost of the raw materials, CO2 and H2, is acceptable. South Africa, which has limited oil reserves, produces 150,000 barrels per day of synthetic liquid fuel using the Fischer-Tropsch process. (13)

An important benefit of synfuels production is that the product is less polluting than fuel obtained by refining crude oil. Figure 7.8 compares a container of conventional diesel fuel refined from crude oil with a container of diesel produced by the Fischer-Tropsch (FT) Process. The FT diesel is much clearer, with much lower concentrations of sulfur and aromatics. (13)

Figure 7.9 shows the reduction in diesel exhaust emissions achieved using FT diesel fuel relative to conventional diesel fuel from crude oil. The reductions in hydrocarbon (HC) emissions, carbon monoxide (CO), carbon dioxide (CO2), nitrogen oxides (NOx) and particulate matter (PM) are substantial.

The FT process and other synfuel production processes starts with Syngas, a mixture of carbon monoxide (CO) and hydrogen (H2) to produce synthetic gasoline and diesel fuel, as described above.

Today, the syngas for the Fischer-Tropsch process can come from a variety of sources, natural gas, coal, biomass, and the source that does not involve fossil fuels or biomass – namely, carbon dioxide from atmosphere air and hydrogen from water.

After extracting carbon dioxide (CO2) from the atmosphere, it would be converted to carbon monoxide (CO) by the reverse water-gas-shift reaction using hydrogen as shown below:

$$CO2 + H2 \rightarrow CO + H2O$$

The reverse water gas shift reaction is simply the old water gas shift reaction in reverse, e.g., the water gas shift reaction is:

$$CO + H2O \rightarrow CO2 + H2$$

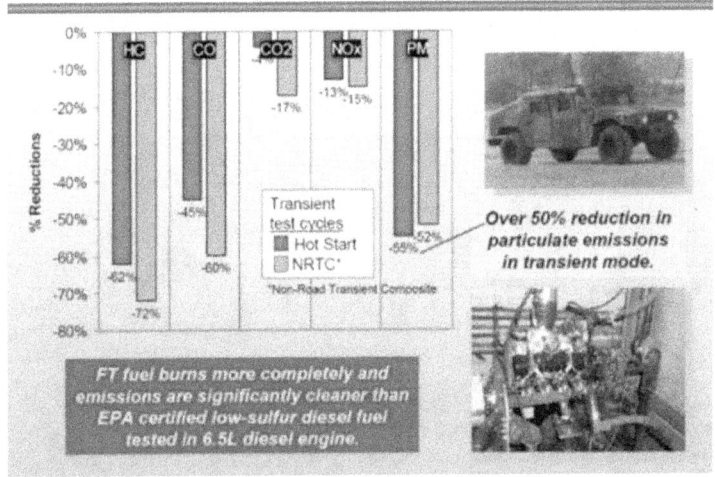

Figure 7.9 OSD Clean Fuel Initiative FT Diesel Emissions Presentation

Depending on reaction condition, one can drive the reaction to make CO from CO2 by supplying hydrogen, or make hydrogen by reacting CO with high temperature steam. The water gas shift reaction was discovered by the Italian physicist Felice Fontana in 1789 and been in commercial use for a hundred years (14).

Combining the Fischer-Tropsch reaction and the reverse water gas shift reaction we have

$$n\,CO_2 \quad + \quad (3n+1)\,H2 \quad \rightarrow \quad C_nH_{2(n+1)} + 2n\,H_2O$$
$$(atmospheric\ air)\ (hydrogen\ fm\ electrolysis) \quad (gasoline)$$
$$(water)$$

Burning a gallon of gasoline releases 19.64 pounds (8.9 kg) of $CO_2$ into the atmosphere. (2) To synthesize gasoline from air and water we extract the same amount of $CO_2$ from the atmosphere 19.64 pounds, as raw material for the reverse Water Gas Shift/Fischer-Tropsch process. The amount of hydrogen feed required per gallon of octane is 2.7 pounds of $H_2$.

The World's air and water inventories are enormous. The effect of withdrawing $CO_2$ from the atmosphere and water from the environment to make synthetic gasoline and diesel will be completely negligible and undetectable.

Figure 7.10 Denali Mt McKinley

The World's atmosphere, a portion of it shown in the beautiful photo of Mount McKinley – now called Denali (Figure 7.10) – has a total mass of 5 million Billion metric tonnes. To obtain the $CO_2$ for 10 Billion barrels of synthetic gasoline and diesel flow, annually, one would extract 10 million tonnes of $CO_2$ from the atmosphere daily, about 12 percent of what we currently emit to the atmosphere each day by burning fossil fuels. To extract the $CO_2$ at its concentration of 400 parts per million, the $CO_2$ extraction plants would take it from about one millionth of the World's atmosphere.

Similarly, electrolyzing water to provide the hydrogen for manufacturing the synthetic gasoline and diesel fuel would have a negligible impact on the World's water resources. Figure 7.11 shows a view of the Mississippi River. Electrolyzing a cubic meter of water produces 110 kg of hydrogen, sufficient to manufacture 92 gallons of synthetic gasoline. For 10 Billion barrels of synthetic gasoline and diesel fuel per year, one would require a water flow of 150 cubic meters per second.

Figure 7.11 View of Mississippi River from Fire Point in effigy Mounds National Monument, Iowa, USA

The average water flow rate of the Mississippi River is 16,792 cubic meters per second (15), more than 100 times greater than required for generating 10 Billion barrels per year. The Amazon is much greater still, with a discharge rate 8 times that of the Mississippi. There are lots of other big rivers around the World from which one would take a very tiny bit of their discharge into the ocean, to generate the synthetic gasoline and diesel. Also, it is likely the synthetic fuel plant will recycle the $H_2O$ produced by the Fischer-Tropsch process and electrolyze it to generate hydrogen. This would reduce the amount of input water for the synfuels plant by 30%.

A variety of processes have been investigated for extracting $CO_2$ from flue gases and the atmosphere. Extraction from the atmosphere is more difficult than from flue gases emitted by fossil fuel fired power plants and other sites, because of the much lower concentration of $CO_2$ in the atmosphere, only 400 parts per million.

$CO_2$ can be scrubbed from the atmosphere by chemical reactions with materials like $CaO$ (quicklime), or an aqueous solution of calcium hydroxide [$CA(OH)_2$], or sodium hydroxide ($NaOH$) or other materials, (16) The $CO_2$ is released from the resultant carbonate and the absorbing material by heating to high temperature. Georgia Tech is working on capturing $CO_2$ from the atmosphere using a ceramic honeycomb structure coated with an absorbent, i.e., a dry amino-modified silica material. The $CO_2$ is removed and the absorbent regenerated by passing steam through the structure. The Georgia Tech researchers estimate that a structure the size of a shipping container – 8x8x20 feet long – could remove 1,000 tons of $CO_2$ per year. To produce 3.7 Billion tons per year for manufacturing 10 Billion barrels of synthetic gasoline/diesel would require a volume equivalent to 3.7 million shipping containers, modest compared to world inventory of 17 million shipping containers. Georgia Tech projected cost for $CO_2$ extracted from the atmosphere is $100 per tonne. (17)

A promising approach for $CO_2$ extraction is the "artificial tree" under development by Klaus Lackner of Columbia University and associates. (18, 19) The artificial tree consists of the multiple plastic sheet surfaces coated with a resin that absorbs $CO_2$ from the air. To recover the $CO_2$, which is then compressed to be used or sequestered, the resin dries in low humidity air to be regenerated.

Lackner estimates that approximately 1.1 megajoules of electrical energy per kilogram of $CO_2$ is required for pumping and compressing the captured $CO_2$ (18). Per tonne of $CO_2$ fed to the Fischer-Tropsch process that corresponds to 1100 megajoules or 305 KWH. At 2 cents per KWH using beamed solar power from space, that's only $6 per tonne energy cost, a small amount. The main cost elements for the artificial tree process will be the capital cost of the structure and absorbent sheets, plus O&M costs. Other estimates project the cost of $CO_2$ removal will be much higher than $100 per tonne, as much as $600 per tonne. (20)

However, at $300 per tonne, synthetic and diesel fuel can be manufactured from air and water at less than $5 per gallon which is less than the cost of today's gasoline and diesel in most countries. $CO_2$ extraction technology is evolving, so that costs will drop substantially. There is general agreement that $CO_2$ can be extracted from flue gases emitted by power plants and industrial processes at $100 per tonne. Initially, $CO_2$ from such sources could be used, reducing the emissions of $CO_2$ from fossil fuels.

How much intake area of atmospheric air into $CO_2$ extraction units will be required to produce 10 Billion barrels per year of synthetic gasoline/diesel? One's first impression is that it must be enormous. However, that's not correct. In fact, the inflow intake are is much smaller than the air flow area through the windmills now operating around the World.

Visualize a 1 m² (10.8 square feet) intake area through which air flows at 10 meters per second (23 mph) to be processed for $CO_2$ extraction. How much $CO_2$ would be retrieved per year? 245 metric tonnes. Per acre of intake flow area, that's, 1 million tonnes of $CO_2$ extracted per year. To manufacture 10 Billion barrels of synthetic gasoline/diesel per year, total intake flow required is only 3700 acres.

To put the intake flow area in a different perspective let's compare it with the swept flow area of windmills (Figure 7.12).

Figure 7.12 Windmill Farm

Shown below are the operating parameter for a typical windmill, the GE 1.5 sle (21).

- 1.5 MW(e) power capacity
- 38.5 meter (126 ft) blade length
- 80 meter (262 ft) hub height
- 118.5 meter (389 ft) total height
- 4,657 m2 (1.15 acre) area swept by blades

Wind mills are BIG structures. The GE 1.5 sle windmill has a total height of 389 feet, 84 feet higher than the torch of the Statue of Liberty (Figure 7.13). It is dwarfed by the 3.0 MW Vestas V 100 windmill (21), which has a total height of 492 feet (150 meters), almost 200 feet higher than the torch on the Statute of Liberty (22).

Total air flow area swept by the blade on a GE 1.5 sle windmill is 4657 m$^2$ (1.15 acre). A $CO_2$ extraction plant with the same intake air flow at 10 meters/sec would produce 1.15 million tonnes of $CO_2$ per year. 3200 such sized air intakes would produce sufficient $CO_2$ for 10 Billion barrels of synthetic gasoline/diesel per year.

Figure 7.13 Statue of Liberty

The $CO_2$ production rate of 245 tonnes per m² per year at 10 meters per second is the maximum possible. It assumes that the time to recover the absorbed $CO_2$ and regenerate the absorbent material is much shorter than the time that the absorbent is exposed to intake air flow. Also, average air flow velocity may be less than 10 meters per second. $CO_2$ removal rate scales as air flow velocity to the 1st power i.e. $(V)^1$, while windmill output power scales as the 3rd power, i.e. $(V)^3$, so that windmills are much more sensitive to wind speed than $CO_2$ extraction units will be. Also, windmills shut down if wind speed drops much below the rated value, while $CO_2$ extraction units will operate over the full range of speeds from zero to maximum rated.

To be conservative, the capacity factor of the $CO_2$ extraction units is taken to be 1/3 (33%) of the 245 tonnes/m² maximum. This allows for down time to recover the absorbed $CO_2$ and regenerate the absorbent and an average wind speed less than 10 meters per second. At 33% capacity factor, 1 m² of airflow intake area would produce 0.33 x 245 = 80 metric tonnes of $CO_2$ per year. A total of 11,000 acres of inflow would be required.

The total area of windmill airflow intake to produce 1.0 Billion barrels per year at 1.15 acres swept area per windmill would then be 3 x 3200 = 9600 equivalent 1.5 MW GE sle windmills. As of the end of 2012, there were over 200,000 windmills operating in the World, with a nameplate capacity of 282,452 megawatts (23). World total wind power generation was 534 terawatt hours, about 4% of the World's total electric power production, with a capacity factor on the order of 40 percent of nameplate rating.

In 2012, the swept area of 282,482 windmills was 282, 482/9600 = 30 times greater than the air intake area to produce 10 Billion barrels of synthetic gasoline/diesel per year. And the number of windmills in the World is rapidly growing. In a few years there will be 50 times as much windmill swept area as needed for $CO_2$ extraction, in a decade or so, 100 times as much swept area.

Having the intake area for $CO_2$ extraction be such a small fraction of the swept area of already operating windmills is very impressive. Also very impressive is the value of the $CO_2$ produced. For the swept area of the GE 1.5 MW sle windmill, 4657 $m^2$, at 80 tonnes per $m^2$ per year and $100 per tonne value, operating for 1 year would yield a $CO_2$ value of 37 million dollars. In comparison, the 1.5 MW(e) turbine, operating at a capacity factor of 40% and an electric value of 4 cents per KWH(e), would produce $210,000 of electric energy annually, a ratio of 180 in value of product.

What will be an acceptable price for synthetic gasoline manufactured from air and water in a non-fossil fuel World economy? It's difficult to predict the future. Absolutely, all we can be certain of is Yogi Berra's prediction – "The future lies ahead."

For synthetic gasoline and diesel from air and water, a target cost of about $5 per gallon appears very reasonable. A large portion of the World pays even more than $5 per gallon today, even in countries with large populations, like India and China.

So, what is the projected cost for synthetic gasoline and diesel from air and water? For simplicity, we project the cost for gasoline, since it is a considerable bigger market than the market for diesel fuel. The cost for diesel will be somewhat higher, but close to the cost for gasoline, perhaps 10 to 20% greater.

The following Table 7.1 summarizes the principal parameters for production of 10 Billion barrels of synthetic gasoline/diesel from air and water using the Fischer-Tropsch process.

Table 7.1
Parameters for Production of Synthetic Gasoline and Diesel Fuel
From Air and Water

| Parameter | Value |
|---|---|
| Barrels of Synthetic Fuel Produced per Year | 10 Billion (420 Billion Gallons) |
| Amount of $CO_2$ extracted from the atmosphere per year | 3.7 x 1 Billion Tonnes, |
| Kilograms of $CO_2$ per gallon | 8.9 Kilograms |
| Amount of $H_2$ Produced by Water electrolysis per year | 500 million tonnes |
| Kilograms of H2 per gallon | 1.2 Kilograms |
| Electrical Energy Per Kilogram of $H_2$ | 50 KwH (80% efficient) |
| $CO_2$ Extraction from Atmosphere Tonnes/year per $M^2$ of intake area at 33% capacity factor | 80 tonnes/year |
| Total CO2 Extraction Air Intake Area for 10 Billion Barrels per Year | 11,000 Acres |
| Swept Air Flow Area for 1.5 MW(e) GE Windmill | 1.15 acre |
| Amount of CO2 Extracted from Atmosphere with Intake Area Equal to Windmill Swept Area at 33% Capacity Factor | 380,000 tonnes per year |
| Total Cost Per Gallon of Fuel for $100 per tonne of $CO_2$ | $3.09 per gallon |
| Total Cost Per Gallon of Fuel for $300 per tonne of $CO_2$ | $4.87 per gallon |

Table 7.2
Price per Gallon of Premium Gasoline Around the World (27)

| Price Range per Gallon ($) | Countries in Price Range (Highest Price Countries First) |
|---|---|
| $10 to $9 | Norway, Denmark, Italy, Netherlands, Greece |
| $9 to $8 | Sweden, Hong Kong, Portugal, U.K., Belgium, France, Finland, Germany, Ireland |
| $8 to $7 | Switzerland, Slovakia, Hungary, Czechoslovakia, Japan, S.Korea, Spain, Slovenia, Austria, Malta, Latvia, Luxembourg, Lithuania, Estonia |
| $7 to $6 | Cyprus, Bulgaria, Australia, Singapore, Romania, Chile, Brazil, India |
| $6 to $5 | Canada, South Africa, Seychelles, Argentina, China |
| $5 to $4 | Thailand, US, Indonesia |
| $4 to $3 | Russia, Malaysia, Mexico |
| $3 to $2 | Iran, Nigeria |
| $2 to $1 | UAE, Egypt |
| $0 to $1 | Kuwait, Saudi Arabia, Venezula |

However, we can get a reasonable idea by looking at what people paid in 2012 for a gallon of premium gasoline in countries around the World. Current prices 2016 are lower due to the much lower cost of oil, a temporary condition. Table 7.2 based on data given by Bloomberg (24). Cost per gallon ranges from $9.69 in Norway to $8.56 in Germany to $7.58 in Japan, to $6.41 in Brazil, to $6.06 in India, to $5.39 in Chia, to $4.19 in the US, to $3.71 in Russia, to 1.79 in Egypt, to $0.61 in Saudi Arabia, to $0.29 in Venezuela. With a crude oil price of $100 per barrel, the crude oil alone costs $2.50 per gallon, plus the cost of refining and shipping. Clearly at a price of less than $4 per gallon, the gasoline is being subsidized.

Since there is no large-scale production plant experience for $CO_2$ extraction from the atmosphere, we consider a range of costs from \$100 to \$300 per metric tonne of $CO_2$. \$100 per tonne is projected by a number of researchers. But on the other hand, some project as high as \$600 per tonne based on present experience. It appears likely that as $CO_2$ extraction technology evolves over the next 10 to 20 years, the cost per tonne should be in the range of \$100 to \$300 per tonne. For 8.9 kg of $CO_2$ per gallon of gasoline, this corresponds to \$0.89 per gallon at \$100/tonne, \$1.78 per gallon at \$200 per tonne, and \$2.67 at \$300 per tonne.

Per gallon of gasoline, the Fischer-Tropsch process requires 1.2 kilograms of hydrogen. A 100 percent efficient electrolysis requires 40 kilowatt hours of electric energy per kilogram of hydrogen produced. At 80% efficiency, a practical value of 50 kilowatt hours per kilogram of $H_2$ is required. Using beamed power from space solar power satellites at 2 cents per KWH, the cost of 1.2 kilograms of hydrogen per gallon of gasoline would be \$1.20.

The 3rd cost component of synthetic gasoline/diesel is the capital cost of the Fischer-Tropsch plant that processes the $CO_2$ and $H_2$ feed materials into gasoline and diesel. Sasol estimates the capital investment for a 96,000 barrels per day FT plant in Westlake, Louisiana to be about 11 Billion dollars (15).

Over a 30-year amortization period, the Sasol plant would produce 96,000 x 365 x 30 = 1.05 Billion barrels. Per barrel the 11 Billion dollar investment corresponds to a cost of \$10.5 per barrel. Per gallon of gasoline that is \$0.25 per gallon for the capital cost of the FT plant.

The 4th cost component is the operating and maintenance costs for the FT plant. A reasonable projection for O&M costs is 10% of the capital investment cost per year, i.e., 1.1 Billion dollars per year. At 96,000 barrels per day, the FT plant would produce $3.5 \times 10^7$ barrels per year. The corresponding O&M cost would then be $1.1 \times 10^9 / 3.5 \times 10^7 = \$31.4$ per barrel or \$0.75 per gallon.

The following Table 7.3 gives the cost of gasoline based on the 4 principal cost elements using the Fischer-Tropsch process with the Reverse Water Gas Shift Reaction.

- Cost of $CO_2$ extracted from the atmosphere
- Cost of $H_2$ produced by electrolysis of water
- Capital cost of the Fischer-Tropsch plant
- Operation and Maintenance costs for the FT plant.

Table 7.3
Projected Cost of Synthetic Gasoline per Gallon from Air and Water
as a Function of Cost per Tonne of CO2 Extracted From Air
Cost per Gallon of Gasoline, in Dollars

| Cost Component | $100/tonne of $CO_2$ | $200/tonne of $CO_2$ | $300/tonne of $CO_2$ |
|---|---|---|---|
| $CO_2$ From Atmosphere | 0.89 | 1.78 | 2.67 |
| $H_2$ From Electrolysis of Water | 1.20 | 1.20 | 1.20 |
| Amortized Capital Cost of FT Process Plant (30 year Amortization) | 0.25 | 0.25 | 0.25 |
| O&M Costs of FT Plant | 0.75 | 0.75 | 0.75 |
| Total Cost Per Gallon | $3.09 | $3.98 | $4.87 |

Table 7.3 shows the total estimated cost per gallon of gasoline, including the above 4 principal cost elements. For $100 per tonne of $CO_2$, total cost per gallon is $3.09, well under the average price paid worldwide. At $200 per tonne, cost is $3.98 per gallon, and at $300 per tonne, cost is $4.87 per gallon, less than most countries already pay for gasoline (Table 7.2).

Conclusion? Manufacturing synthetic gasoline and diesel fuel from $CO_2$ and $H_2$ is nothing new. The Fischer-Tropsch and other processes have been operating for many decades at large scale. Currently, 260,000 barrels of synthetic fuel are manufactured daily at acceptable prices. Manufacturing large amounts of hydrogen at acceptable cost by electrolysis of water is not new either, and in fact, is the basis for the hydrogen fuel cell automobile program.

The new element for the proposed program for synthetic fuel generated from air and water is the extraction of $CO_2$ from the atmosphere as feed material for the synthesis. Various processes for $CO_2$ extraction have been tested and found feasible but as yet, no large scale plants have operated. Estimates of the cost of extracting $CO_2$ range from $100 per tonne to much higher values, on the order of $600 per tonne.

Depending on the cost of the extracted $CO_2$, the estimated cost of synthetic gasoline from air and water on the order of $3 per gallon at $100 per tonne of $CO_2$ and $5 per gallon at $300 per tonne of $CO_2$. The current price of gasoline derived from oil is more than $5 per gallon in most countries around the World, so that $5 per gallon for synthetic gasoline from air and water would be acceptable for most countries.

As a first step, $CO_2$ emitted from fossil fuel fired power and industrial plants, for example, the cement industry, which accounts for about 5% of global $CO_2$ emissions, could be used as feed material for the synfuels process as the technology for $CO_2$ extraction develops and cost figures become more precise. Because of the higher $CO_2$ concentrations in emissions from fossil fuel fired power plants and industries, the cost per tonne of $CO_2$ is relatively low, about $100 per tonne at maximum. As $CO_2$ extraction from air technology evolves, and fossil fuels are phased out, the synthetic gasoline and diesel fuel plants would transition to $CO_2$ extraction from the atmosphere.

By 2050 AD, it appears possible to have a mix of Maglev, electric cars, and synthetic gasoline/diesel fuel for the World's transport needs, with no need for fossil fuels, and no net $CO_2$ emissions into the atmosphere.

Audi, the car maker, has already started manufacturing diesel fuel from air and water at the Sunfire pilot plant at a capacity of 42 gallons per day, in Dresden, Germany.

Their projected costs are 1 to 1.2 Euros per liter corresponding to $4 to $5 dollars per gallon using electricity at present prices. With lower-cost power beamed from space the price of synfuels manufactured from air and water would be even less.

The details of the work at Audi can be found at:

http://money.cnn.com/2015/04/28/technology/audi-diesel-air-water/

http://www.gizmag.com/audi-creates-e-diesel-from-co2/37130/

# Part III

# Implementing Maglev to Stop Global Warming

*"My Lords, During the short time I recently passed in Nottinghamshire not twelve hours elapsed without some fresh act of violence, ...I was informed that forty Frames had been broken the preceding evening. These machines ... superseded the necessity of employing a number of workmen, who were left in consequence to starve. By the adoption of one species of Frame, in particular, one man performed the work of many, and the superfluous labourers were thrown out of employment....The rejected workmen in the blindness of their ignorance, instead of rejoicing these improvements in art so beneficial to mankind, conceived themselves to be sacrificed to improvement in mechanism."*

With these words Lord Byron in his maiden speech to the House of Lords in February, 1812, recognized the nexus of economics and politics in explaining the revival of the Luddite protest that was shaking the English social order. The "frames" were the new machines that employers were installing in the workshops of England's growing textile industry. At the time of the speech, the Parliament had before it legislation to exact the death penalty for such acts of sabotage.

History has sustained that technological invention beginning with the invention of the steam engine, internal combustion engines and electricity powered with fossil fuels and successive waves of technological innovation have brought spectacular growth in both employment and real wages, a combination that spells prosperity and a higher quality of living.

However, as *Silent Earth* explains, it is an urgent imperative that our collective investment of the World's capital must concentrate on shifting this prosperity to new, sustainable sources of energy.

The shift won't be easy for the reasons cited by Lord Byron. Because there is a natural fear that the manufacturing, mining, and transport industries of the modern world, which are dependent on energy generated by fossil fuels must be superseded and will result in unemployment and a reduction in the standard of living.

In the 1812 English parliament, the Earl of Lauderdale sharpened Lord Byron's thesis that the misled workers encountered by Byron in Nottinghamshire were acting against their own interests: *"Nothing could be more certain that the fact that every improvement in machinery contributed to toe improvement in the conditions of persons manufacturing the machines, there being in a very short time after such improvements were introduced a greater demand for labour than ever before."*

Fortunately, the technologies and basic inventions that can make the shift from fossil hydrocarbons practical and feasible are already here and only need to be developed. Their development will create new industries to evolve transport and the sources of much more efficient the generation of electricity. We only must overcome the resistance of existing capital investment and the fears of people that a shift from fossil hydrocarbons will reduce their standard of living. This is the historical role of governments who must reach a political consensus and lead in the demonstration and safety of these technologies to attract the private investment that has proven successful in economic history.

*Silent Earth* will be published in 2016 prior to the US Presidential elections in November. This election could and should be a turning point on minimizing the emissions of global warming gases that will cause a large-scale and potentially catastrophic change in climate that will extend longer than the entire history of human civilization, thus far. This is the finding of a growing number of scientific papers that reflect the opinion of the majority of the Earth and climate scientists.

Policy decisions made during this political and policy window of opportunity over the next few years are of utmost importance to the well-being and survival of humanity.

# Chapter 8

# The Economics of Implementing Maglev

To be an attractive approach for helping to eliminate combustion of fossil fuels, the Maglev applications described in chapters 4-7 – surface transport, low-cost electricity beamed from Space Solar Power Satellites, low-cost energy storage, and production of synthetic liquid hydrocarbon fuel from air and water – must have:

- Affordable implementation costs
- Major economic savings in the present costs of passenger and freight transport, electrical generation, energy storage, and synthetic fuel production
- Major savings in the present human costs of transport accidents and health damage from air pollution.
- Capability to operate as a network, connecting all population centers, with convenient access to most of the population.
- Capability to transport passengers, autos, trucks, and freight.

First, what are the costs for implementing global Maglev transport, and how do they compare to World GDP, expressed as Gross World Product (GWP) in the period 2015 to 2050 AD?

Historically, in the 20th Century, the annual rate of growth of World GWP was in the range of 3 to 4 percent. In constant 1990 US dollars, World GWP was 1.1 Trillion dollars in 1900 rising to 41 Trillion dollars in 2,000 AD (1), as illustrated in Table 8.1, an average growth rate of 3.6% per year.

Annual World GWP at 2000 AD was 37 times greater than the annual World GWP at 1900 AD. Similar growth rates can be expected in the 21st Century.

## Table 8.1

Annual Gross World Product (GWP) from 1900 to 2000 AD

In Constant 1990 US Dollars (1)

For an average annual GWP growth rate of 3.8%

| Year | Real GWP ($ billions, 1990) | Compound annual growth rate |
|---|---|---|
| 2000 AD | 41,016.69 | 4.04% |
| 1990 AD | 27,539.57 | 4.14% |
| 1980 AD | 18,818.46 | 4.43% |
| 1970 AD | 12,137.94 | 5.87% |
| 1960 AD | 6,855.25 | 4.77% |
| 1950 AD | 4,081.81 | 3.12% |
| 1940 AD | 3,001.36 | 2.91% |
| 1930 AD | 2,253.81 | 1.4% |
| 1920 AD | 1,733.67 | 2.29% |
| 1900 AD | 1,102.96 | 2.69% |

In the time period 2005 to 2014, Gross World Product (GWP) has grown at the following rates: (2)

Annual GWP growth rate (%)

| | 2006 | 2007 | 2008 | 2009 | 2010 | 2011 | 2012 | 2013 | 2014 | 2015 (est.) |
|---|---|---|---|---|---|---|---|---|---|---|
| World Average | 5.5 | 5.7 | 3.1 | 0 | 5.4 | 4.2 | 3.4 | 3.4 | 3.4 | 3.5 |

Price Waterhouse Coopers has projected future GDP's from 2014 to 2050 AD for the 20 largest economies in the World. (3)

Table 8.2 shows the projected GDP's for the top 5 countries, for 2014, 2020, 2030, and 2050.

Table 8.2

Projected Annual GDP's for the 5 Largest Economies in the World, for 2014, 2020, 2030, and 2050 AD.

In Trillion of 2014 US Dollars (3)

| Rank | 2014 | 2020 | 2030 | 2050 |
|------|------|------|------|------|
| 1. | US (17.4) | US (20.3) | China (26.7) | China (53.6) |
| 2. | China (10.4) | China (15.9) | US (25.4) | US (41.4) |
| 3. | Japan (4.8) | Japan (5.2) | India (7.3) | India (28.0) |
| 4. | Germany (3.8) | Germany (4.2) | Japan (6.0) | Indonesia (8.7) |
| 5. | France (2.9) | United Kingdom (3.3) | Germany (4.7) | Brazil (8.5) |
| Total GDP Top 5 Countries | 39.3 | 48.8 | 70.2 | 140.2 |
| Total GDP Remaining 15 Countries | 22.8 | 27.1 | 38.9 | 74.3 |
| Total GDP 20 Countries | 62.1 | 75.9 | 109.1 | 214.5 |
| Total GWP for World | 77.9 | 94.9 | 136.4 | 268.1 |
| Total Integrated World GWP 2015 to 2050AD | | | | 5,380 Trillion 2014 USD |

Total GWP in 2014 for the World was 77.9 Trillion in US Dollars. Total GDP for the 20 largest economies in Table 8.2 was 62.1 Trillion 2014 USD, corresponding to the ratio, 20 largest GDP/Total World GWP =62.1/77.9=80%. That is, the 20 largest economies accounted for 80% of total Gross World Product. The same ratio of 80% has been used to calculate Total World Product (GWP) in 2020, 2030, and 2050 AD. That is, total GWP in these years is calculated by dividing the sum of the 20 largest countries for the given year by 0.80.

Projections of future GDP's and GWP's are not certain, of course, and events may change the numbers. However, the projections appear reasonable. The average annual growth rate for the World GWP from 2015 to 2050 AD is 3.5%, which is in the middle of the historical range of 3 to 4 percent over long periods.

The integral of annual World GWP over the period 2014 to 2050, that is, the average annual GWP times 35 years is very large, 5,380 Trillion dollars. The average annual GWP over the 35 years period in 154 Trillion dollars, twice the 2014 annual GWP of 78 Trillion dollars, and 58% of what the annual GWP will be in 2050.

How does the cost of the Global Maglev Transport Network described in Chapter 4 compare with the total World GWP for the 35 years from 2014 to 2050 AD? It is tiny.

The following Table 8.3 compares the projected investment in the Global Maglev Network described in Chapter 4 with the total integrated Gross World Products of 5,380 Trillion 2014 USD and the World total capital investment in conventional land transport infrastructure of 45 Trillion 2014 USD from 2014 to 2050, projected by the International Energy Agency (IEA) [4]. Adding in the reconstructing upgrade, and operating and maintenance costs for transport infrastructure, the total is 120 Trillion dollars, 2.2% of total World GWP.

Total capital cost of the guideways for the Pan American, Pan African, and Pan Europe-Asia Maglev Networks is 2.8 Trillion 2014 US Dollars. Adding in the capital costs of the Maglev stations and vehicles, plus the local transport infrastructure that would provide access to the many convenient Maglev stations on the Maglev Networks plus 100% contingency, the total Capital Investment for the Global Maglev Network is 15 Trillion 2014 USD, only 0.3% of the total World GWP from 2014 to 2050. The Maglev investment is small, 1/3, of the IEA projected Capital Investment in road and rail systems to 2050 AD.

Table 8.3
Projected Investment in the Global Maglev Network
From 2020 to 2050

| Investment Compared | Investment Amount (2014 USD) |
|---|---|
| **Guideway Investment** | |
| Pan-American Network | 1.1 Trillion |
| Pan-African Network | 0.5 Trillion |
| Pan Europe/Asia Network | 1.2 Trillion |
| Maglev Stations & Vehicles | 2 Trillion |
| Transport Infrastructure, Providing Access to Maglev Stations | 3 Trillion |
| 100% Contingency | 7 Trillion |
| Total Maglev Capital Investment | 14.8 Trillion |
| Maglev Capital Investment As Fraction of $5,380 Trillion 2014 USD Gross World Product | 0.28% |
| IEA Estimate of Capital Investment for Conventional Land Transport. | 45 Trillion |

*Global Land Transport Requirements, Estimating Road and Railway
Infrastructure Capacity and Costs to 2050, IEA, John Dulac
https:www.iea.org/.../Transport Infrastructure

Maglev transport systems require much less operating and maintenance that conventional road and rail systems, because there is no mechanical contact, steel-wheel pounding on steel rails, and rolling friction. Local loading of structures due to levitation carries from Maglev vehicles is 100s of times smaller than the loads imposed on the wheels of motor vehicles – particularly, heavy trucks – and railroad cars, including road and rail O&M and upgrades to 2050 AD, IEA projects a total of 120 Trillion 2014 USD, 3 times greater than the 45 Trillion 2014 USD Capital Investment. Maglev O&M will be much smaller. Assuming that Maglev O&M is equal to the 15 Trillion dollars capital investment, which is probably more than its actual cost, Maglev total cost, capital plus O&M would be 30 Trillion 2014 USD, 1/4th of the IEA estimate of 120 Trillion 2014 USD for road and rail systems.

Accordingly, implementing Maglev will save many 10's of Trillions of dollars that would otherwise be spent on the construction & maintenance of conventional road and rail systems in the World, during the period from 2015 to 2050.

There will be even greater savings for passengers and freight from Maglev's lower transport cost for passengers and freight compared to conventional highway, rail and air modes. As discussed in Chapter 4, freight on highway trucks transported by Maglev is projected to cost 10 cents per ton mile, compared to 30 cents a ton mile by highway. Passenger travel on Maglev vehicles will cost 3 cents per passenger mile, compared to 15 cents per mile by air, and 40 cents per mile by highway.

Expenditures for transportation are a substantial fraction of national GDP for all countries, particularly the developed countries. US Bureau of Labor Statistics for 2009 estimate the percentage of national GDP spent on transportation – personal, public, and logistics of goods – for the following countries to be (5):

| Country | % of National GDP |
| --- | --- |
| US | 18% |
| Canada | 20% |
| UK | 15% |
| Japan | 10% |

The fraction spent on transportation for the US and Canada is higher than for the United Kingdom and Japan, because relative consumer expenditures for automobiles are higher for the US and Canada compared to the UK and Japan. US and Canada are big countries with much lower population densities than Japan and the UK.

A survey of 17 developed countries' transport spending in 2005 as a share of national GDP, carried out by Dr. Jean-Paul Rodrique at Hofstra University (6) found that their transport shares were in the range of 10 to 18%, with the US being the highest. This is consistent with Figure 7.3 in Chapter 7, which shows the higher per capita GDP in a given country, the higher the number of automobiles per 1000 people.

The statistics of transport's share of GDP for less developed countries are not available, but are probably smaller than for developed countries. However, less developed countries are rapidly growing in GDP, particularly China and India, which have 1/3 of total World population.

Expressed in 2014 US Dollars, China's GDP is projected to increase from 10.4 Trillion dollars in 2014 to 53.6 Trillion dollars in 2050 AD, a factor of 5 increase. India is projected to increase from 2.0 Trillion dollars in 2014 to 28.0 Trillion dollars in 2050, a factor of 14 increase.

Total World GWP is projected to increase from 78 Trillion dollars in 2014 to 268 Trillion dollars in 2050, in constant 2014 US Dollars, a factor of 3.5 increase. World population is projected to increase from 7 Billion in 2014 to 9 Billion in 2050, a factor of only 1.3 increase. Assuming that these estimates prove accurate, the average World GDP per capita, at $11,000 per person in 2014 will increase to $30,000 per capital in 2050.

The result? When income per capita increases, more money is spent on transportation. People travel more, buy more automobiles, and buy more goods, more varied foods, which have to be transported from their source to their distribution.

This increase in future transport is shown in the projections of the International Energy Agency (IEA). In the IEA's 2013 report, Global Land Transport Infrastructure Requirements (7) the following global transport estimates are given (Table 8.4).

Table 8.4
Estimated Global Travel Growth in the $4^0$ C DS Scenarios (7)

| Transport Mode | 2014 | 2050 |
|---|---|---|
| Passenger Light Duty Motor Vehicles | 16 Trillion Passenger Km | 42 Trillion Passenger Km |
| Airplane Travel | 2 Trillion passenger Km | 10 Trillion Passenger Km |
| Public Transit | 9 Trillion Passenger Km | 15 Trillion Passenger Km |
| Road Freight | 4 Trillion Tonne Km | 18 Trillion Tonne/Km |
| Rail Freight | 10 Trillion Tonne Km | 25 Trillion Tonne KM |

Particularly noteworthy are the large increases in passenger motor vehicle and road freight transport. The above increases correlate with the projected increase in World GWP, as expected.

Total GWP increases by a factor of approximately 3 from 2014 to 2015, while total transport, passenger kilometers plus freight tonne kilometer, also increases by approximately 3, from 40 Trillion passenger kilometers plus freight tonne kilometers in 2014, to 110 Trillion passenger kilometers plus freight tonne kilometers in 2050.

A reasonable, and probably conservative estimate for the transport share of total Word GWP is 10%. For developed countries, the transport share is somewhat higher, for less developed countries, it is probably somewhat lower.

At 10% transport share, based on existing road, rail, ship, and air transport technologies, the total cumulative World expenditures on transport from 2014 to 2050 would be:

| World Transport Expenditures | = (0.10) X Word GWP from 2014 to 2050 |
|---|---|
| 2014 to 2050 AD | = (0.10) 5380 Trillion Dollars (2014 USD) |
| | = 540 Trillion Dollars (2014 USD) |

Implementation of Maglev will substantially reduce the projected transport cost of 540 Trillion dollars for conventional transport, because of Maglev's much lower cost per passenger kilometer or passenger mile, whichever unit is preferred, and much lower cost per truck tonne-kilometer, or ton-mile, as preferred.

We project that with large scale implementation of Maglev, total World transport costs as a share of World GWP could be reduced from 10% with conventional transport systems down to 5% with Maglev. Reduction to 5% would save 270 Trillion dollars.

There would be substantial additional savings enabled by Maglev implementation resulting from reductions in highway accidents and health damage to people's hearts and lungs from the pollutants and microparticles emitted by motor vehicles.

It is difficult to estimate these addition savings. In the US, the National Highway Traffic Safety Administration estimated that in 2010, the economic cost of highway accidents was 877 Billion dollars, 6 percent of US GDP (8). The additional health damage effects would also be very large.

What would be the World-wide economic savings from reduced fewer accidents and less health damage when Maglev is implemented? To be conservative, we project a 2 percent savings in World GWP, about 100 Trillion dollars.

Total transport, accident, and health savings from implementing Maglev? 370 Trillion dollars, a factor of 25 greater than the projected 15 Trillion dollars for global implementation.

And there are additional major savings from implementing Maglev launch to put space solar power systems in orbit that would beam down to Earth clean, very low cost electrical power.

Total World electrical generation in 2014 was 22,700 Terawatt Hours (9). One Terawatt Hour equals 1 Billion Kilowatt Hours. Electrical generation in 2050 is projected to be considerably higher, on the order of 50,000 Terawatt Hours, a factor of more than 2 greater.

Table 8.5
Levelized Cost of US Electrical Generation as Function of Mode of Generation

| Generation Mode | Levelized Generation $/MWt |
|---|---|
| Conventional Coal | $95 |
| Advanced Coal | $116 |
| Advanced Coal with Carbon Capture | $144 |
| Natural Gas-Conv. Combined Cycle | $75 |
| Advanced Combined Cycle | $73 |
| Adv. Comb. Cycle w/Carbon Capture | $100 |
| Advanced Gas Turbine | $113 |
| Advanced Nuclear | $95 |
| On-Shore Wind | $74 |
| Off-Shore Wind | $197 |
| Solar PV | $125 |
| Solar Thermal | $240 |

The levelized cost of electrical generation depends on the method of electrical generation, Table 8.5 shows values of the levelized generation cost for different modes for new electrical generation in 2020 (10). Values are shown in US dollars per Megawatt Hour (MWH).

For the US, the average generation cost is on the order of $100 per Megawatt Hour, equivalent to 10 cents per Kilowatt Hour.

Annual World expenditures for 22.7 Terawatt Hours in 2014 at $100 per MWH correspond to 2.3 Trillion dollars, about 3% of World GWP. In 2050 AD, with 50,000 Terawatt Hours, annual World electrical generation expenditures would be 5 Trillion dollars, approximately 2% of GWP.

At an average expenditure of 2.5% of World GWP per the 2014 to 2050 period, the total amount spent for electrical generation using conventional modes of generation would be 0.025 x 5380, equal to 135 Trillion dollars at 8 cents per KWH:.

The cost of beamed power from space solar power satellites launched into orbit by Maglev is analyzed in Chapter 5. The estimated cost is $20 per MWH (2 cents per KWH). The net savings using beamed space solar power is $80 per MWH, 8 Cents per KWH.

Over the period from 2014 to 2050, the net global savings in the cost of electric generation that would occur if beamed space power were available would be.

Net electric power savings = 0.8 x135 Trillion Dollars
= 108 Trillion Dollars (2014 USD

Adding together the 2014 to 2050 savings in the costs of transport, highway accident and health damage and electrical generation, total savings would be:

| Transport Savings | 270 Trillion Dollars (2014 USD) |
|---|---|
| Avoided Highway Accidents and Health damage | 100 Trillion Dollars (2014 USD) |
| Electrical Generation | 108 Trillion Dollars (2014 USD) |
| Total Savings | 478 Trillion Dollars (2014 USD) |

Clearly, the above estimates arc not precise. However, it appears reasonable that the net savings enabled by implementing Maglev could be on the order of 500 Trillion dollars (2014 USD) or roughly 10 percent to the total of about 5,000 Trillion dollars (2014 USD) projected for the total World GWP from 2014 to 2050.

The savings enabled by Maglev could be smaller or greater than the estimates shown above. Prophets and oracles don't always get things right, as noted in Chapter 9. However, it appears very likely that the savings from implementing Maglev will be considerably greater than the 15 Trillion dollars cost to implement it.

# Reveille
by A. E. Housman (1859-1936)

*Wake: the silver dusk returning*
*Up the beach of darkness brims,*
*And the ship of sunrise burning*
*Strands upon the eastern rims.*
*Wake: the vaulted shadow shatters,*
*Trampled to the floor it spanned,*
*And the tent of night in tatters*
*Straws the sky-pavilioned land.*
*Up, lad, up, 'tis late for lying:*
*Hear the drums of morning play;*
*Hark, the empty highways crying*
*'Who'll beyond the hills away?'*
*Towns and countries woo together,*
*Forelands beacon, belfries call;*
*Never lad that trod on leather*
*Lived to feast his heart with all.*
*Up, lad: thews that lie and cumber*
*Sunlit pallets never thrive;*
*Morns abed and daylight slumber*
*Were not meant for man alive.*
*Clay lies still, but blood's a rover;*
*Breath's a ware that will not keep.*
*Up, lad: when the journey's over*
*There'll be time enough to sleep.*

# Chapter 9

# The Politics of Implementing Maglev

To successfully implement a radically new technology that disrupts major existing transport, electrical generation, and fuel supply systems, 3 conditions must be met:

- The new technology must be practical to construct and operate reliably and safely.
- The new technology must offer major economic benefits, both to consumers and companies that construct and operate it.
- The new technology must have strong political support to proceed with its implementation even though there will be strong political opposition from the various interests that are vested in the existing competitive systems.

Superconducting Maglev technology has been proven. The 1st Generation System, now operating in Yamanashi, Japan, has carried 100,000 passengers at speeds up to 370 mph, with an accumulated running distance of several hundred thousand miles. The 1st Generation System has operated safely and reliably. Japan Rail plans to extend the Yamanashi Maglev line to become a 300 mile route between Tokyo and Osaka that will transport 100,000 passenger miles daily with a trip time of 1 hour.

Maglev technology is evolving, as previous transport technologies have done. Airplanes evolved from the small, slower, propeller driven passenger planes in the 1930s, e.g., the DC-3 and the Ford Tri-Motor planes, to today's much faster, bigger, and cheaper fare modern jet liners like the Boeing 767 and Airbus. Autos have evolved from the Ford Model T to today's multitude of comfortable and faster cars. Trains have evolved from slow steam driven locomotives to electrically powered 200 mph trains like France's TGV.

The Maglev 2000 System is the 2nd generation of Maglev transport systems, with the new and unique capability to carry trucks and autos as well as passengers. It can also use existing railroad tracks that have been adapted for Maglev travel at very low cost. While not yet operational, the full scale 2nd generation system components – superconducting magnets for Maglev vehicles, aluminum loop guideway panels, a monorail guideway beam, and a passenger vehicle fuselage have been successfully fabricated and tested. With further funding, 2nd Generation Maglev components can be incorporated into vehicles that would demonstrate high-speed operations on low-cost elevated monorail guideways and the capability for levitated travel on existing railroad tracks.

Operating in evacuated tunnels or tubes at low pressure, the absence of air drag of Maglev vehicles gives the StarTram Maglev Launch System the capability to electrically launch payloads into space without requiring rockets.

Chemical rocket launch costs are on the order of $5,000 per kilogram of payload launched into Low Earth Orbit, and considerably higher for launch into Geosynchronous Orbit for beaming space solar power down to Earth. StarTram can launch payloads into orbit at much lower cost, on the order of $50 per kilogram a factor of 100 cheaper, enabling low cost space solar power to be beamed down to Earth.

The basic technology for Maglev Launch already exists – Maglev levitation and propulsion of heavy vehicles to high speeds, tunnels of the length required for launch, large volume vacuum systems, and protective coating and cooling methods for ICBM re-entry vehicles that travel down from space to ground targets at speeds comparable with Maglev launch vehicles coasting up from the ground to space.

Maglev launch to space, while an engineering challenge, is much less challenging than the Manhattan and Apollo projects which started from a much less developed technology base and achieved success in a very short time, only a few years.

The Maglev projects described in Part 2 of *Silent Earth* are practical, and can be brought to full fruition in the next 20 years if pursued with vigor and adequately funded. Technology will not limit the implementation of Maglev as a way to help stop global warming.

Similarly, economics will not limit the implementation of Maglev. As described in Chapter 8, there are enormous economic benefits that will result from implementing global Maglev transport, Maglev launch to space of solar power satellites that beam very low cost electric power down to Earth, and maglev energy storage.

The construction costs of the Maglev systems described in Part 2 are tiny, much less than 1% compared to the Gross World Product (GWP), estimated to total more than $5,000 Trillion (in 2014 US dollars), over the next 35 years to 2050 AD. The economic savings from implementing Maglev in terms of much lower costs for passengers and freight transport, electrical power generation, and synthetic fuel production will be on the order of 10's of Trillions of dollars, much more than the Maglev construction costs.

In addition, there also will be tremendous benefits in terms of much fewer deaths and injuries from highway accidents and their economic costs, as well as greatly reduced lung and heart health damage from pollution and microparticles presently emitted by motor vehicles and merchant ships. Plus, faster travel for passengers and freight and reduced congestion delays, will substantially increase economic productivity and efficiency.

The big barrier to implementing Maglev will not be the productivity of its technology, nor its economics, which will yield tremendous benefits. The big barrier will be political.

Implementation of the Maglev systems described in the book will be very disruptive to many powerful industrial and commercial interests, and to the persons who profit from and are employed by them. They will strongly oppose Maglev implementation, and exert tremendous pressure on politicians and policy makers to not implement the systems proposed here.

Opposition will come from the corporations that extract fossil fuels, oil, coal and natural gas, the companies that refine the fuels, and the many individuals and companies that market to consumers, e.g. gas stations, electric power companies, auto, truck, and airplane manufacturers, and airlines, trucking companies and individual truckers, automobile dealers, highway construction companies and workers, etc., etc.

Rachel Carson's book, *"Silent Spring"*, published in 1962 aroused tremendous opposition from the chemical industry because it threatened their production of agricultural pesticides. She was viciously attacked by the industry and their lobbyists, threatened with lawsuits, called a communist sympathizer, etc.

However, in the end, she prevailed. Following her death in 1963 DDT was banned, the Environmental Protection Agency (EPA) was established, and a strong environmental movement was born in America.

Why did she succeed? Basically, for 3 reasons. First, the industrial and political opposition to her was only a very small part of the America economy. The impact of banning pesticides had only a tiny effect on livelihoods of the American public, both personally and economically.

Second, people love birds. Who can be against them? People want them around. It's very pleasant to hear them singing and see their beautiful shapes and colors. No more birds? No way!

Third, the impacts of pesticides on bird populations could be felt in the near term. To save them, one must act quickly. It's not like global warming, where deniers can say there's no problem, it's just normal natural variations, or if there may be a problem, we can wait a few decades before we have to act. It's human nature to put-off acting on some problem that's many years down the road – we'll get to it someday, right now we have these near term problems.

So, how do we build support for implementing Maglev as the way to help stop global warming and prevent mass extinction?

The public must become more aware of the need to stop consuming fossil fuels as quickly as possible, if humanity is to avoid global catastrophe. However, the effort to communicate to and motivate the public to action has been going on for many years, with minimal real results.

Policy makers talk the talk, but don't walk the walk. Meeting are held, goals set for carbon dioxide reduction, promises are made, and protocols are signed, but global carbon dioxide emissions keep rising every year. For the most part, the public is not aware of, denies the existence of, or thinks that they don't have to worry about it now, it is too hard to worry about the coming catastrophe from global warming. As a result, the opposition from vested industrial and commercial interests prevents any real action – carbon tax, megatrends, funding new technologies, etc. – from happening.

A more effective approach than trying to build public support for reducing fossil fuel consumption by articles, books, speeches, from those who recognize the coming environmental catastrophe is to push for new non-fossil fuel technologies that bring major benefits in the short term.

Maglev ground transport and space launch are such technologies. Shorter trip times, lower cost, more convenient, and comfortable service, safer travel -- who would not choose Maglev over highway, air, and rail?

3 hours, New York to Chicago instead of 16 hours by highway, at lower cost than driving with no chance of accidents? And you can deliver 5 times as much ton-miles per truck in your fleet at lower cost per ton-mile than going by highway. And you don't have to find long distance drivers and risk accidents. The choice for Maglev is easy, simple, and convenient.

Once 2nd Generation Maglev systems begin implementation there will be tremendous public support and ridership for them, Maglev routes will quickly expand into the national and global networks described in Chapter 4.

What will drive the development and implementation of 2nd Generation Maglev? Where will the initial funding come from? While there will certainly be strong opposition to the development and initial implementation of 2nd Generation Maglev by many existing industrial and commercial interests, the possibility of investment in Maglev by some of the existing interests should not be ruled out.

Figure 9.1

Mene, Mene, Tekel, Uphsrasin. The handwriting on the Wall in Belshazzars Palace in ancient Babylon (Figure 9.1), depicted by Rembrandt, forecasting the destruction of Babylon by the Medo-Persians and the death of Belshazzar.

Well, it all came to pass. Goodbye, Babylon. Not all industrial and commercial interests that presently control transport and space launch are so short sighted that they do not see the Handwriting on the Wall for fossil fuels. And they realize that investing now in the coming new technologies is vital to their economic future.

And there are major investment opportunities for companies and individuals that are not involved in present fuel, transport and space launch systems. Getting in on the ground floor with a modest investment in Maglev, which can become the World leader in ground transport and space launch systems would yield enormous economic returns.

And the investment field is not limited to just one nation, if America refuses to invest in Maglev, its policy since 1992, when the House Transportation Committee refused to hold hearings on Senator Moynihan's $750 million dollar legislation for an American Maglev R&D Program. His legislation passed the Senate. Had it passed the House committee, chaired by Representative Robert Carr from Detroit, America would now have an operating 29,000 mile National Maglev Network transporting passengers, trucks, and autos.

If America does not act to develop, implement, and market 2$^{nd}$ Generation Maglev, some other country will and reap the enormous economic manufacturing benefits, exporting it all around the World including the US.

## In Conclusion: Will Humanity Give Up Fossil Fuels and Avoid Global Catastrophe?

If humanity does not give up fossil fuels there will be global catastrophe – mass extinction of species, ocean acidification, and the collapse of modern civilization. That seems certain. Whether humanity will give up fossil fuels or not, that is not certain.

Prophesying the future is hard. We can take the approach of the great Yankee philosopher, Yogi Berra, who said, "The future lies ahead". That's not much help, however.

We can seek the help of Sibyls and Oracles. They were very popular in ancient times, and still are today, though they are now labeled as pundits, analysts, hedge funds, pollsters, etc. In general, however, their predictive power is not much better than the ancient Sibyls and Oracles.

Michelangelo's painting of the Delphic Sibyl (Figure 9.2) portrays her as the Granddaughter of Poseidon, the God of the Ocean. She made prophecies in the Precinct of Apollo, on the slopes of Mr. Parnassus. According to legend, she lived through 9 generations of men. After death, she continued as a voice that foretold the future in dark riddles. That's neat, but was she accurate? A bit more real was the Oracle of Delphi. Located in the Temple of Apollo at Delphi, the resident priestess, named Pythia, would chew oleander leaves and inhale their smoke. Then she would utter prophecies that could be interpreted in different ways. In 560 BC, Croesus, King of Lydia, asked the Oracle what would happen if he invaded neighboring Persia. The Prophecy? "A Great Empire would be destroyed". Croesus took the prophecy to mean Persia would be destroyed. Unfortunately, it was his Empire that was destroyed.

One would think that rational persons would not rely on the Delphic Oracle. However, the leaders of various nations, Sparta, Macedon, Persia, Athens, and Thera, gladly sought guidance from the Oracle – just like today.

Then there was Cassandra, Daughter of King Priam and Queen Hecuba of Troy. Cassandra (Figure 9.3) was blessed by Apollo with the gift of prophecy. Then Apollo tried to have sex with her and she refused. Angered over her ingratitude, Apollo spat in her mouth and cursed her – she would prophesize accurately but nobody would believe her. She foresaw the abduction of Helen by Paris, the Trojan War, the Greeks hiding in the Trojan Horse, and the destruction of Troy. Nobody believed her.

After the fall of Troy she was raped by Ajax from the Greek Army. The goddess Athena angered by the rape, had Zeus and Poseidon cause storms that wiped out most of the Greek fleet as they sailed back home after the Fall of Troy. Cassandra then was taken as a concubine by King Agamemnon of Mycenae, only to be murdered along with Agamemnon by his Queen, Clytemnestra and her lover Aegis. Thus, Cassandra did end-up in the Elysian Fields after her death judged by the Gods to be a worthy person.

Things are a bit better career-wise for today's Sybils and prophets, thankfully. However, will they be anymore able to really predict whether humanity will give up fossil fuels in time to avoid global catastrophe than the ancient Sybils and prophets accurately predicted the future? We certainly can't, and it's very unlikely that anybody can.

258

| Figure 9.2 | Figure 9.3 |

So, we're back to Yogi Berra – "The Future Lies Ahead." We can't predict it, but what could it look like? Hieronymus Bosch's wondrous triptych, reproduced on back cover, can give a glimpse of our possible future.

The left panel, that shows Adam and Eve emerging into Eden, can be taken as a symbolic rendering of the emergence of humanity into the beautiful and pristine natural world.

The central panel can be taken as a symbol of the World today. Eden is no more – humans have created their own World, some parts good, some bad, some prosper and are happy, others not. Nature is something in the background, largely ignored.

The right panel depicts Hell, with universal suffering, pain, and unhappiness. One can take that as symbolic of the terrible World that humanity will experience if it does not give up fossil fuels.

So do we give up fossil fuels and switch to clean energy services that can sustain modern society long-term, or do we keep consuming fossil fuels and go the Hell in the next few decades? We don't know. Nobody knows. We do know that humanity can avoid Hell if it chooses. Whether it will or not, who knows?

# Acknowledgements

The development of the advanced 2nd Generation Maglev 2000 System would not be possible without the following wonderful people who have worked tirelessly with great dedication to bring Superconducting Maglev to America.

The authors wish to express our deep gratitude and thanks to our following colleagues in Superconducting Maglev. Listed alphabetically, they include:

Cliff Bragdon; Gil Carpenter; Bob Coullahan, Paul Earle, Ernie Fazio, Les Finch; Bud Gardner; Bud Griffis, Ed Harmer; John Jackson; Piyush Joshi; Chris Kempner; Otto Lazareth; George Maise, John Morena; Erno Ostheimer; Judy Otto; Jim Paul; P. Philipsburg; Sal Pjerov; Jesse Powell; John Rather; Douglas Rike; Rosemary Riker; Morris Reich; Barbara Roland; Art Shenfeldt; John Skaritka; Charles Smith; Richie Tulipano; Bill Van Allen; Lou Ventre; Tom Wagner; Brian Walsh; Cindi Weavers; Rod Wickers; John Wines; Bob Worthing; Al Witzig.

We also wish to express our deep gratitude and thanks to the memory of the late Senator Daniel Patrick Moynihan, the champion for American Maglev. Had his Senate legislation for Maglev been signed into law 25 years ago, Americans would now be riding on fast comfortable vehicles and shipping and consuming goods delivered on the National Maglev Network.

The research and fact-checking for the book was greatly helped by the Internet, Wikipedia, and Google Maps. We also appreciate the wonderful people who work in the Senate library, who kept a watchful eye out for publications relevant to the environmental, energy, and economic issues that impact the deployment of Maglev Technology. We are especially grateful to Tamara Elliott. She commutes by rail, and quickly saw the benefits of superconducting Maglev. Her ability to find publications of interest was unfailing.

We also give our appreciation to the teachers and students from the Middle Schools on Long Island, who participate in the annual Maglev contest at Brookhaven National Laboratory, featuring as guests, Dr. Powell and Dr. Danby. The enthusiasm of the students and teachers make us very proud. Finally, the graphics and illustrations in this book, which are crucial to the reader's understanding, would not have been possible without the able skills of Mr. Douglas Rike.

Senator Daniel Patrick Moynihan's death on March 26, 2003, at age 76, was announced on the Senate floor by then Senator Hillary Rodham Clinton, who was elected to the Senate seat Moynihan had held for 24 years. She said *"We have lost a great American, an extraordinary senator, an intellectual and a man of passion and understanding for what really makes the country work."*

Known for his ability to spot emerging issues and trends, Moynihan was a leader in welfare reform and transportation safety policy and an authority on Social Security and foreign policy.

Senate Democratic leader Tom Daschle of South Dakota said *"in many respects, Pat Moynihan was larger than life,"* citing a description in the Almanac of American Politics that Moynihan was *"the nation's best thinker among politicians since Lincoln and its best politician among thinkers since Jefferson."*

The Maglev Story would not be complete without honoring my colleague, Dr. Gordon Danby. We have worked together on Maglev since the idea was first conceived on the road to Boston in 1959, resulting in the presentation of our paper to the American Society of Mechanical Engineers in 1966, 50 years ago. Without Gordon's deep understanding of the basic physics of magnetism, Maglev 2000 would not have happened.

Gordon Danby is widely respected for his contribution to the practical application of theoretical science to technology. His achievements are recognized by his peers as changing Magnetic Resonance Imaging and Transportation Industries.

From the Franklin Award citation, "Danby's pioneering research efforts in magnetic technology led to the production of open Magnetic Resonance Imaging (MRI) machines that are better, faster and more patient friendly than their tunnel-style predecessors. Danby, along with James Powell, also invented the Superconducting Maglev, a magnetically levitated, high speed train system. The practical and efficient design of the Maglev provides mixed freight and passenger service and interfaces easily with other transport modes."

Dr. Danby received his B.S. in physics and math from Carleton University in Ottawa, Canada, and his Ph.D. in nuclear physics from McGill University in Montreal, Canada. He is a fellow of the American Physical Society. In 1983, the New York Academy of Sciences honored Danby with the Boris Pregel Award for Applied Science and Technology

.

# Appendix

## Silent Earth References

### Prologue

1    Life Expectancy, Our World in Data, ourworldindata.org/data/population-growth-vital/../life-expectancy/

2    List of Regions by Past GDP(PPP) Per Capita, https://en.wikipedia.org/wiki/List_of_regions_by_past_GDP_(PPP)_per_capita

3    World Population, https:en.wikipedia.org/wiki/world_population

4    Fossil Fuel, https://en.wikipedia.org/wiki/fossil_fuel

### Chapter 1

1    2013 GDP (nominal) per capita, https://en.wikipedia.org/wiki/List_of_countries_by_ GDP (nominal)_per_capita, United Nations 2013 Data

2    http://data.worldbank.org/indicator/EG.USE.PCAP.KG.OE

3    Life Expectancy_Our World In Data, our world in data. org/data/population—growth—vital/.../life-expectancy/

### Chapter 2

1    BP Statistical Review of World Energy June 2015 bp.com/statistical review

4    Global Warming, https://en.wikipedia.org/wiki/global_warming

5    www.CSIRO.au/greenhouse-gases/

6    Climate Change:Vital Signs of the Planet:Evidence http://climate.nasa.gov/evidence/

7    The Arctic Climate Threat That Nobody's Even Talking About Yet, Chris Mooney, Washington Post April 1, 2015

8    Methane Clathrate, https://en.wikipedia.org/wiki/methane_clathrate

## Chapter 3

1. Hardin, Garett, The Tragedy of the Commons, Science, 162, p.1243-1248 (1968)

2. https://en.wikipedia.org/wiki/carbon_dioxide

3. Carson, Rachel, Silent Spring, Houghton Mifflin (1962)

4. Silent Spring, https://en.wikipedia.org/wiki/Silent_Spring99

5. https;//en.wikipedia.org/wiki/Greenland_ice_sheet

6. http://www.bbc.com/news/science-environment 28852980

7. https://en.wikipedia.org/wiki/Antarctic_ice_sheet.

8. https://en.wikipedia.org/wiki/sea_level_rise

9. https://en.wikipedia.org/wiki/retreat_of_glaciers_since_1850

10. http://www.epa.gov/climate_change/impacts--adaptation/energy. html

11. https://en.wikipedia.org/wiki/1970_Bhola_cyclone

12. https://en.wikipedia.org/wiki/hurricane_Sandy

13. https://en.wikipedia.org/wiki/climate_change_and _agriculture

14. https://en.wikipedia.org/wiki/2003_European_heat_wave

15. https://en.wikipedia.org/wiki/chicago_heat_wave

16. https://en.wikipedia.org/wiki/Effects_of_global_warming_on_human_health

17. https://en.wikipedia.org/wiki/mountain_pine_beetle

18. https://en.wikipedia.org/wiki/ocean_acidification

19. http: time for changes.org/ocean-acidification-risk-from-global-warming.

20. https://en.wikipedia.org/wiki/extinction

21. https://en.wikipedia.org/wiki/extinction_risk_from_global_warming

22. https://en.wikipedia.org/wiki/The_Raft_of_the _Medusa

## Chapter 4

1. www.eia.gov/biofuels/issues/trends

2. Corn Production in the United States, https://en.wikipedia.org/wiki/corn_production_in _the United _States

3. Transport Outlook 2011 Meeting the Needs of 9 Billion People OECD/ITF 2011

4. Global Greenhouse Gas Emissions Data, http:www.epa/gov/climate change/ghgemissions/global.html

5. List of countries by GDP(PPP) per capita,en.wikipedia.org/wiki/list_of_countries_by_GDP_(PPP)_per_capita

6   List of countries by vehicles per capita,
    https://en.wikipedia.org/wiki/list_of_countries_by_vehicles_per_capita

7   List of Countries by Past and Future Population,
    https://en.wikipedia.org/wiki/List
    _of_countries_by_past_and_future_population

8   List of IMF Ranked Countries by Past and projected GDP(nominal),
    https://en.wikipedia.org/wiki/list_of countries_by_future_GDP

9   Trans-Siberian Railway, https://en.wikipedia.org/wiki/Trans_Siberian_Railway

10  List of Countries by Rail Transport Network size
    https://en.wikipedia.org/wiki/list_of_countries_by_rail_transport_network_siz
    e

11  Rail Transport in South Africa,
    https://en.wikipedia.org/wiki/rail_transport_in_South_Africa

12  Narrow Gauge Railways in Africa,
    https://en.wikipedia.org/wiki/narrow_gauge_railways_in_Africa

13  Trans-Asian Railway, https://en.wikipedia.org/wiki/Iron_Silk_Road

14  International Shipping and Word Trade=Facts and Figures IMO Library
    Services, February, 2006

15  https://en.wikipedia.org/wiki/answers.com/Q/How_much_fuel_does_a_conta
    iner_ship_burn

16  Big Polluters, one massive container ship equals 50 million cars, Paul Evans,
    http: www.gizmag.com/shipping -pollution/1526

17  US warns of pollution from merchant ships off the Florida Coast,
    www.guardian.co/UK/environment/2009/March/31/noaa-pollution-
    Florida_freighters_tankers_cruise_ships

## Chapter 5

1   StarTram, The New Race to Space, James Powell, George Maise, and Charles
    Pelligrino, Amazon.com

2   Apollo Program, en.wikipedia.org/wiki/Apollo_Program

3   International Space Station,
    en.wikipedia.org/wiki/International_Space_Station

4   Furon Corporation, "Space Transportation Costs: Trends in Price per Pound to
    Orbit: 1990 to 2000", September 6, 2002

5   Saturn V, en-wikipedia.org/wiki/Saturn_V

6   Space Shuttle, en-wikipedia.org/wiki/Space_Shuttle

7   The Case for Space Solar Power, John C. Mankins, Virginia Edition Publishing,
    LLC (2014)

8     List of Countries by Electricity Production, en-wikipedia.org/wiki/list_of_countries_by_electricity_production

9     International Energy Outlook 2013, www.eia/gov/forcast/ieo/electricity.cfm

10    Power from the Sun, Its Future, Peter E. Glaser, Science Magazine 62(3856), 857-861(November 22, 1968)

11    Space Based Solar Power, en.wikipedia.org/wiki/space_solar_ power

12    Cost of Electricity by Source, en-wikipedia.org/wiki/cost_of_electricity_by_source

13    List of IMF Ranked Countries by Past and Projected GDP (nominal), en.wikipedia.org/wiki/list_of_countries_by_future_GDP

## Chapter 6

1     Existing Capacity by Energy Source, http://www.eia.gov/electricity/annual/html/epa_04_03.html

2     Pumped Storage Hydroelectricity,

3     http://en.wikipedia.org/wiki/Pumped_storage_hydroelectricity

4     Electric Power Annual 2007: A Summary, http://www.eia.gov/electricity/annual

5     http: large.stanford.edu/courses/2012/ph240/doshay/docs/034808.pdf

6     Powell, J. and Danby, G. 1966 High Speed Transport by Magnetically Suspended Trains Paper 66-WA/RR-5 American Society of Mechanical Engineers, NY, NY. Also, 1967, A 350 mph Magnetically Suspended Train, Mech.Eng. 89, 30-35

7     Danby, G., Jackson, J., and Powell, J. 1974 Calculations for Hybrid (Ferro-Null Flux) Low Drag Systems, IEEE Transactions on Magnetics May 10, p.443-446

8     International Energy Outlook 2010, US EIA, http://www.eia.gov/forecasts/archive/aeo10/index.html

## Chapter 7

1     Carbon Dioxide, en.wikipedia.org/wiki/Carbon_Dioxide

2     Carbon Dioxide From Gasoline and Diesel Fuels

3     http://www.eia.gov/facqs/fag.cfm?id=307Bpt=10

4     Total World CO2 Emissions, http://www.iea.org/newsroom and events/news/2012/may/name

5     US energy-related Carbon Dioxide Emissions by Sector, 2011, http://www.eia.gov/energy _in_brief/images/charts/energy_relations

6    Energy in the United States en.wikipedia.org/wiki/energy_in_the_United States

7    Short Term Energy Outlook, US Energy Information Administration, http://www.eia.gov/forecasts/steo/report/global

8    Transportation and Energy, Dr. Jean Paul Rodrique and Dr. Claude Contois, https//people.hofstra.edu/geotrans/eng/ch8en/conc:en?ch8c2en.html

9    List of IMF Ranked Countries by Past and Projected GDP (nominal) http;//en.wikipedia.org/wiki/list_of_countries_by_future_GDP

10   http://www.eia.gov/toos/faqs/fag.ctm?id=24 et+10

11   History of the Electric Car,

12    http://en.wikipedia.org.wiki/history_of_the_electric_car

13   Electric Car, en.wikipedia.org./wiki/electric_car

14   Fischer-Tropsch process-en.wikipedia.org/wiki/Fischer_Tropsch_Process

15   Synthetic Fuel, en.wikipedia.org/wiki/synthetic_gasoline

16   Water-gas shift reaction, en.wikipedia.org/wiki/water_gas_shift_reaction

17   Mississippi River, en.wikipedia.org/wiki/Mississippi_river

18   Carbon Dioxide Removal, en.wikipedia.org/wiki/carbon_dioxide_removal

19   Technology Developed for Extracting Carbon Dioxide from Air, efficiently Maria Reyes:

20    http://www.green optimist.com/2012/07/25/technology_developed

21   400 ppm: can artificial trees help pull CO2 from the Air?, David Biello, http://www.scientific American.com/article/prospects-for-direct-air...(May 16, 2013

22   Sucking CO2 from the skies with artificial trees, Gala Vinee, (October 4, 2012), http://www.bbc.com/future/story/c0121004-fake-trees-to-clean-...

23   Direct Removal of Carbon Dioxide From Air Likely Not Viable, Report suggests, Science Daily (May 9, 2011), http://www.sciencedaily.com/releases/2011/05/110509114200.htm

24   Technical Specs of Common Wind Turbine Models(AWEO.org), http:awes.org/windmodels.html

25   Statue of Liberty, en.wikipedia.org/wiki/statue_of_liberty

26   Wind Power, en.wikipedia.org/wiki/wind_power

27   Highest and cheapest Gas Prices by Country, Mark J Perry (May 15,2012) Wall Street pit.com/02107_highest_and_cheapest_gas_prices

# Chapter 8

1   J. Bradford DeLong (24 May 1998) "Estimating World GDP, One Million BC-Present", https:// en.wikipedia.org/wiki/Gross_World_Product

2   IMF Data, GWP Growth Rate, https:// en.wikipedia.org/wiki/Gross_World_Product

3   List of IMF Ranked Countries by Past and Projected GDP (nominal), https:// en.wikipedia.org/wiki/List_of_IMF_ranked_countries_by_past_and_projecte d_GDP (nominal)

4   Global Land Transport Requirements, Estimating Road and Railway Infrastructure Capacity and Costs to 2050, IEA, John Dulac

5   www.bls.gov/opub/focus/volume2/_number 16/cex_2_16.htm

6   Transport Spending as Share of GDP, Selected Countries, https://people.hofstra.edu/geotrans/.../transport_gdp.ht

7   Figure 1, Global Land Transport Requirements, IEA, John Dulac

8   The Economic and Societal Impact of Motor Vehicles Crashes, 2010, National Highway Traffic Safety Administration, DOT HS 812013(May 2014)

9   World Energy Consumption, https:// en.wikipedia.org/wiki/World_Energy_Consumption

10  Annual Energy Outlook 2015, EIA, www.eia.gov/forecasts/aeo/electricity_generation.cfm

# List of Figures and Credits

## Dedication

- Photo of Rachel Carson: https://commons.wikipedia.org/wiki/File:Rachel_Carson.jpg Author, Fish and Wildlife Service

## Prologue

- Figure 1 https://commons.wikimedia.org/wiki/File:Population_curve.svg. Author, EIT

## Part I

- Figure 2 https://commons.wikimedia.org/wiki/paradise_lost_12.jpg. Author, Gustav Dore

## Chapter 1

- Figure 1.1 Gross Domestic Product per Capita as Function of Energy Use Per Capita for Countries, Author, J.Powell

- Figure 1.2   US Constitution:https:/commons.wikimedia.org/wiki/File:US_ Constitution_1997.svg. Author, Todd Stevens; Maschee Spazierganger https:/commons.wikimedia.org/wiki/File:Maschee_Spazierganger.jpg. Author, AxelHH; London-Farrington Coach passing Buckland House, Berkshire https://commons.wikimedia.org/wiki/File:Jamess _Pollard ...., Author, James Pollard; The Cow Boy 1888 – https://commons.wikimedia.org/wiki/File: The _Cow_Boy_1888.jpg. Author, John C.H. Grabill

- Figure 1.3 Mount Fujijapan:https:/commons.wikimedia.org/wiki/File: MountFuijapan.jpg, Author, Swallito; Virgin Atlantic: https://commons.wikimedia.org/wiki/File:Virgin.atlayntic.a340-600.g- Vyou.argo.jpg., Author, Adrian Pingstone; 2011 Toyota Corolla: https://commons.wikimedia.org/wiki/file:2011Toyota_Corolla. Author, US National Highway Traffic Safety Administration; Freedom of the Seas: https:commons.wikimedia.org/wiki/File:MS_freedom_of_the_seas ..... Author, Andres Manuel Rodriquez

- Figure 1.4 Riders Pony Express. https://commons.wikimeda.org/wiki/file:riders_Pony_Express.jpg Author: Ernest and Elaine Hartnagle; New England Courant: https://en.wikimedia.org/wiki/file: New England Courant_00001.jpg. Author, unknown; Pony Express Poster: https://commons.wikimedia.org/wiki/file:Pony_Express_Poster.jpg. Author, Pony Express

- Figure 1.5
- The Great Train Robbery: https://commons/wikimedia.org/wiki/file:the_great_train_robbery.jpg Author, Edwin S. Porter
- Mobile Phone Evolution: https:commons.wikimedia.org/wiki/File: Mobile_phone_evolution.jpg
- Cptv Display.https://en.commons.wikimedia.org/wiki/File:Cptvdisplay.jpg. Author, Wags 05
- Laptop:https://commons.wikimedia.org/File: Laptop.jpg. Author, Jon Sullivan

- Figure 1.6 Historical Time Line of Life Expectancy, J.Powell

- Figure 1.7 Life Expectancy with Fossil Fuels, Author, J.Powell Reference 1: 2013 GDP (nominal) per capita; https: en.wikipedia.org/wiki/List_of_countries_by_GDI-capta United Nations (2013) Reference 2: List of countries by Life Expectancy; https://en.wikipedia.org/wiki/List_of_countries_by 2013 World Health Organization (WHO) Data

- Figure 1.8
- Wheat Pennsylvania 1943: https://commons.wikimedia.org/wiki/File: Wheat Pennsylvania 1943.jpg, Author, John Collier
- Poultry of the World: https://commons.wikimedia.org/wiki/File: Poultry_of_the_World.jpg. Author, L.Prang&Co.
- Market Vegetables: https://commons.wikimedia.org/wiki/File:marketvegetables.jpg, Author, Jasper Greek Golangco
- Hereford bull: https://commons.wikimedia.org/wiki/File:Hereford _bull_large.jpg. Author, Robert Merkel

- Figure 1.9
- Supermarket Checkout:https://commons.wikimedia, org.wiki/File: Supermarket_check_out.jpg. Author, Velela
- First-Kitchen Fast Food: https://commons.wikimedia.org/wiki/File: First_Kitchen_Fast_Food.jpg, Author, Kici
- Fridge, interior: https://commons.wikimedia.org/wiki/File. Fridge interior.jpg Authors, Seasons greetings
- Tin Cans-Three: https://commons.wikimedia.org/wiki/File:Tin Cans_Three.jpg. Author, Seth Ilys

- Figure 1.10
- Schenk-Crooke House: https://commons.wikimedia.org/wiki/File: Schenk_Crooke_House2.jpg, Author, unknown
- Mount Vernon View: https://commons.wikimedia.org/wiki/File: Mount_Vernon_with_the_Washington_family_on_the_terrace.jpg, Author, Benjamin Henry Latrobe.
- Philadephia President's House: https://commons.wikimedia.org/wiki/File:Philadelphia Presidents House.jpg, Author, William L. Breton
- Saskatchewan sod house: https://commons.wikimedia.org/wiki/File: Saskatchewan_sod_house.jpg, Author, Mrs. Ed Brusseau

- Figure 1.11
- Ames, Iowa, Mobile Homes Underwater: https://commons.wikimedia.org/wiki/File:Fema-44989_Ames, Iowa_Mobile_Homes_Under_water.jpg. Author, Leo Jace Anderson
- Cornelius Vanderbuilt House, https://commons.wikimedia.org/wiki/File:Cornelius_Vanderbilt_House_II.jpg, Author, Bain News Service.
- Red Brick Flats London, https://commons.wikimedia.org/wiki/File://red.birck.flats.london.arp.jpg, Author, Adrian Pingstone
- New Ranch White, https://commons.wikimedia.org/wiki/File:
- New Ranch White.jpg, Author, Mcheath

- Figure 1.12
- Young Man with a Candle, https: commons.wikimedia.org/wiki/File:Gobin_Michel_young_man_with_a_candle.jpg , Author, Michel Gobin
- Franklin Stove, https:commons.wikimedia.org/wiki/file:Franklin_stove.jpg, Author, Unknown
- Fireplace Burning, https:commons.wikimedia.org/wiki/file: Fireplace_burning, Author, Francisco Belard
- James Peale by Charles Wilson Peale, https:commons.wikimedia.org/wiki/file:James_Peale_by_Charles_Wilson_Peale.jpg, Author, Charles Wilson Peale

- Figure 1.13
- Constructing heated floor, https://commons.wikimedia.org/wiki/File:constructing heated floor.jpg, Author, Rpvdt
- Baywater Power Station with coal, https://commons.wikimedia.org/wiki/File: Baywater_Power_Station_with_coal.jpg, Author, webaware
- View of US from Space at Night, New York City at Night  Author NASA-NOAA

- www.nasa.gov https://commons.wikimedia.org/wiki/File:New_York City_at_Night.jpg, Author, NASA

## Chapter 2

- Figure 2.1 Lithograph of Manhattan Island, http://commons Wikipedia.org/wiki/File:George_Schlegal_George_Degan_New_York_1873, jpg, Authors, George Schlegal (artist) and George Degan (publisher)

- Figure 2.2 Twin Towers_NYC, http://commons.wikipedia.org/wiki/file:twintowers -NYC.jpg, Author, Carol M. Highsmith

- Figure 2.3 Cumulative Height of Fossil Fuels by World if they were laid on top of Manhattan Island, Author, J. Powell

- Figure 2.4 Years in which Fossil Fuel Reserves are Exhausted for 3 Different Consumption Scenarios, Author, Powell

- Figure 2.5 Mauna Loa CO2 Monthly Mean Concentration, https://commons.wikimedia.org/wiki/File:Mauna_Loa_CO2_monthly_mean_con centration.jpg Author, Delorme

- Figure 2.6 Atmospheric Carbon Dioxide From Law Dome Ice Cores, source, CSIRO

- Figure 2.7 Historical Evidence of CO2 in Earth's Atmosphere, https://commons, Wikimedia.org/wiki/File: Evidence_CO2. Jpg Author, NASA

- Figure 2.8 The Greenhouse Effect, https://www.ncdc.noaa.gov/paleo/globalwarming/whal.html, Author, NOAA

- Figure 2.9 Absorptivity of Thermal Radiation in Air as a Funcdtion of Carbon Dioxide Pressure and Radiation Path Length, Source, Perry's Chemical Engineering Handbook, 6th ed, McGraw Hill (1984)

- Figure 2.10 Vostok Data: https://commons.wikimedia.org/wiki/file:Vostok_Petik_data.jpg Author, NOAA.

- Figure 2.11 Energy Change Inventory, 1971-2010, https://commons.wikimedia.org/wiki/File: Energy_change_inventory, 1971-2010. Jpg, Author, Enescot

- Figure 2.12 Photo of Arctic Permafrost, https://www.flickr.com/photos/USFWS_Alaska/6757752485/sizes/e, Author, US Fish and Wildlife Service, Alaska

- Figure 2.13 Photo of Arctic Permafrost Canyon, https://www.flickr.com/photos/US Geological Survey/12116729703, USGS

- Figure 2.14 Photo of Burning Methane Hydrate, https://commons.wikimedia.org/wiki/File:Burning_hydrate_inlay_US_Office_Nav al_Research.jpg, Author, US Office of Naval Research

# Chapter 3

- Figure 3.1 Photo of Lacanja Forest Burn, https://commons.wikimedia.org/wiki/File: Lacanja_burn.jpg. Author, Jami Dwyer

- Figure 3.2 Photo of Midtown Manhattan from Gantry Plaza, https:commons.wikimedia.org/wiki/File: Midtown Manhattan From Gantry Plaza.jpg. Author, Igo Iric

- Figure 3.3 Photo of Rachel Carson, https://commons.wikimedia.org/wikimedia.org/wiki/file:Rachel_Carson.jpg. Author, Fish and Wildlife Service

- Figure 3.4 Map of Arctic Year Long Temperature Anomaly, https: commons.wikimedia.org/wiki/File:Arctic Yearlong Temp Anom Hr.jpg. Author, Hunter Allen and Richard Rivera, NOAA

- Figure 3.5 Photos of Melt Ponds on Arctic Sea Ice, https://commons.wikimedia.org/wiki/file: Ponds_on_the_Oceans_ICESCAPE.jpg. Author, NASA

- Figure 3.6 Arctic Sea Ice Minimum Comparison, https://commons.wikimedia.org/wiki/File: Arctic_Sea_Ice_Miknimum_comparison.jpg. Author, NASA

- Figure 3.7 Map of Greenland Albedo Change, https://commons.wikimedia.org/wiki/File: Greenland_Albedo_Change.png. Author, NOAA

- Figure 3.8 Satellite view of Antarctica, https:commons.wikimedia.org/wiki/File:Antarctica_6400px_from_Blue_Marble.jpg, Author, NASA

- Figure 3.9A Photo of Iceberg in North Star Bay, Greenland, https://commons.org/wiki/File: Iceberg_in_North_Star_Bay_Greenland.jpg. Author, Jeremy Harkbeck, NASA

- Figure 3.9B Photo of the Edge of the Ross Ice Shelf in Antarctica, https://commons.wikimedia.org/wiki/File: Corp2400_Flickr_NOAA_Photo Library.jpg. Author, Nathaniel B. Palmer

- Figure 3.10A Photo of Whitechuck Glacier, 1973, https: commons.wikimedia.org/wiki/File:Whitechuck_glacier_1973.jpg. Author, Mauri Pelto

- Figure 3.10B Photo of Whitechuck Glacier, 2006, https: commons.wikimedia.org/wiki/File: Whitechuck_glacier_2006.jpg. Author, Peltoms

- Figure 3.11 Trends in global average absolute sea level, https://commons.wikimedia.org/wiki/file:trends_in_global_average_absolute_sea_level_1880-2013.png Author, USEPA

275

- Figure 3.12 6 meter Sea Level Rise, https:commons.wikimedia.org/wiki/file: 6m_sea_level_rise.jpg, Author, NASA

- Figure 3.13 EPA Map of Florida Areas Flooded by 1,3,and 6 meter Sea Level Rise, and Effect on Power Plants, Https://www.epa.gov/climate change/impacts-adaptation/energy.html. Author, EPA

- Figure 3.14 Satellite Photo of Hurricane Katrina, https://commons.org/wiki/File: Hurricane-Katrina-Aug 28-05-2145UTC.jpg. Author, NOAA

- Figure 3.15 Photo of Katrina New Orleans Flooded; https://commons.wiimedia.org/wiki/File:Katrina/New Orleans Flooded Edit2.jpg. Author, Kyle Niemi, US Coast Guard

- Figure 3.16 Photo of Katrina Bayou LaBatre.2005 boats ashore, https://commons.wikimedia.org/wiki/File:Bayou_La_Batre_2005_ashore.jpg. Author, NOAA

- Figure 3.17 Photo of Hurricane Ike Gilchrist damage, Https://commons.wikimedia.org/wiki/file: Hurricane_Ike_Gilchrist_damage.jpg. Author, Jocelyn Augustine/FEMA

- Figure 3.18 Photo of Bolivar 62(IMG 9193), https://commons.org/wiki/File: Bolivar 62(IMG_9193).jpg. Author, National Weather Service

- Figure 3.19 Satellite Photo of Hurricane Sandy Oct 28, 2012, https://commons.wikimedia.org.wiki/File:Sandy_Oct_28_2012_15557.jpg. Author, NASA

- Figure 3.20 Photo of Hurricane Sandy damage Long Beach Island, https:commons.Wikimedia.org/wiki/file:Hurricane_Sandy_damage_Long_Beach_Island.jpg. Author, The National Guard

- Figure 3.21 Map of California Drought Status, https:commons.wikimedia.org/wiki/File:California_Drought_status_October 21_2014.png. Author, Michael Brewer, NCDC/NOAA

- Figure 3.22 Map of West U.S. Drought, https://drought monitor.Unl.edu/home/regional drought monitor.as...Author, Anthony Artusa, NOAA/NWS/NCEP/CPE

- Figure 3.23 Photo of President Obama and Governor Brown at California Farm affected by drought, https://www.whitehouse.gov/sites/default/files/image_file. Author, US White House

- Figure 3.24 Satellite Photo of California Wildfires, https://commons.wikimedia.org/wiki/File:May_2014_California_wildfires_close-up.jpg. Author, NASA

- Figure 3.25 Satellite Photo of California Fires, https://commons.wikimedia.org/wiki/File:California_Fires-MODIS081715.jpg. Author, NASA

- Figure 3.26 Photo of California Wildfire, https://commons.wikimedia.org/wiki/File:wildfire_in_California.jpg. Author, Bureau of Land Management

- Figure 3.27 Photo of Forest Fire Aftermath, https://commons Wikimedia.org/wiki/File:Forest_fire_afermath.jpg. Author, Beasterline at the English project

- Figure 3.28 NASA Map of Wildfire Intensity in the Western US and Canada, https://svs.gsfc.nasa.gov/vis/a060000/a004000/a004092/USCentered Fire Frequency FRP40.jpg. Author NASA

- Figue 3.29 Projectred changes in yields of selected crops with global warming, https://wikimedia.org/wiki/File: Projected_changes_in_yields_of_selected_crops_with_global_warming.jpg. Author, Enescot

- Figure 3.30 Map of Temperature Rises in Europe During 2003. Heat Wave, https://commons.wikimedia.org/wiki/File:Canicule_Europe_2003.jpg. Author, Reto Stockli and Robert Simmon

- Figure 3.31 Pine Beetle, https://commons.wikimedia.org/wiki/File:Pendroctonus_ponderosae.jpg. Author, US Forest Service

- Figure 3.32 Gala Apple Branch with scorched leaves after fire blight infection File: Severe fire blight infection on apples.jpg- Wikimedia commons. Author, Peggy Greb, US Dept of Agriculture

- Figure 3.33 Graph of 5 Previous Mass Extinctions on Earth, https://en.wikipedia.org/will/File:Extinction intensity.jpg. Author, Rursus

- Figure 3.34 Painting of the Extinct Dodo, https://commons.wikimedia.org/wiki/File:Edward's _Dodo.jpg. Author, Roelant Savery

- Figure 3.35 Painting of the Passenger Pigeon, https://commons.wikimedia.org/wiki/File:Ectopisles_migratorius NCN2P28CA.jpg. Author, Hayashi and Toda (artists) Charles Otis Whitman (Author)

- Figure 3.36 Photo of the Golden Toad, https://commons.wikimedia.org/wiki/File:Bufo_periglenes 2.jpg. Author, Charles H. Smith

- Figure 3.37 The Siren, oil on canvas, Leeds Art Gallery

  Edward Armitage - http://images.bridgeman.co.uk/cgi-bin/bridgemanImage.cgi/600.LMG.0816210.7055475/123001.JPG

- Figure 3.38 The Raft of the Medusa is an oil painting of 1818–1819 by the French Romantic painter and lithographer Théodore Géricault. Completed when the artist was 27, the work has become an icon of French Romanticism.

# Chapter 4

- Figure 4.1 Photo of 1s Generation Japanese Superconducting Maglev. Author, Japan Railway

- Figure 4.2 Drawing of Roll-on, Roll-off Maglev Vehicles Carrier Author, D. Rike

- Figure 4.3 Drawing of Maglev Auto Carrier Author, Douglas Rike

- Figure 4.4 Map of the Long Island Railroad, web.met.info/lirr/timetable, lirr, map.htm, Author, MTA

- Figure 4.5 Drawing of Interstate Highway with Maglev Guideway Author, Douglas Rike

- Figure 4.6 Average Daily Long Haul Freight Traffic on the National Highway System, 2002 &2035, Author, USDOT

- Figure 4.7 Accident scene on a US Highway www.nrd.dot.gov/pubs/811402.pdf, Author, USDOT

- Figure 4.8 Congestion on a US Highway, Author, USDOT

- Figure 4.9 Map of the US National Maglev Network, Authors, Powell and Jordan

- Figure 4.10 First Maglev Wave to be completed 10 years from Start of US Maglev Program, Authors, Powell and Jordan

- Figure 4.11 Map of the West Coast Maglev Network, Authors, Powell and Jordan

- Figure 4.12 Map of the NorthEast – Midwest Porthion of the East Coast Maglev Network, Authors, Powell and Jordan

- Figure 4.13 Map of the Second Wave of the US National Maglev Network, Authors, Powell and Jordan

- Figure 4.14 Map of the Third Wave of the US National Maglev Network, Authors, Powell and Jordan

- Figure 4.15 Photo of New York City Subway Car, https/en.wikipedia.org/wiki/New_York_City_Subway, Author, Unknown

- Figure 4.16 Map of New York City Subway System Adapted to Maglev, Authors, Powell and Jordan

- Figure 4.17 New York City Subway Track Adapted for Maglev Service Author, Hynchui, Polytechnic University of Brooklyn

- Figure 4.18 Map of the World, http://en.wikipedia.org/wiki/Earthmap. 1000x500compc.jpg, Author, unknown

- Figure 4.19 Satellite view of the Strait of Gilbralter, https//en.wikipedia.org/wiki/Strait_of_Gilbralter_crossing, Author, NASA

- Figure 4.20 The Trans-Siberian Railway Map, http://commons.wikimedia.org/wiki/File: Map_ Trans_Siberian_Railway, Author, User. Stefan Kuhn

- Figure 4.21 Satellite View of the Bering Strait, http://en.wikipedia.org/wiki/Bering_Strait, Author, NASA

- Figre 4.22 Map of PanAmericanHighway, http://en.wikipedia.org/wiki/Pan_American_Highway, Author, en.user.Seeweege

- Figure 4.23 Map of Railway Network in China, http://en.wikipedia.org/wiki/Rail_transport_in_China, Author, Alancrh

- Figure 4.24 Map of India's Transport Networks, www.cia.gov, Author, CIA

- Figure 4.25 Automobiles Per 1000 persons as a Function of GDP per Capita for Different Countries, Authors, Powell and Jordan

- Figure 4.26 Map of Pan African Maglev System, Authors, Powell and Jordan

- Figure 4.27A, Map of Trans-Asian Railway (TAR) Network, https://sites.google.com/site?Indian ocean community/trans-asia-railway, Author, United Nations

- Figure 4.27B, Map of Pan Europe-Asia Maglev System, Authors, Powell and Jordan

- Figure 4.28 Hanjin Container Ship in San Francisco Bay, Wikipedia, Author, Unknown

## Chapter 5

- Figure 5.1 The Apollo 11 Saturn V Vehicles Liftoff, en.wikipedia.org/wiki/File:KSC_69pe-442.jpg, Author, NASA

- Figure 5.2 The Space Shuttle Discovery (STS 120) Lift off for the International Space Station, en.wikipedia.org.wiki/file:STS 120 Launch, HIRes-Edit.jpg, Author, NASA

- Figure 5.3 Ascent Trajectory to LEO Using StarTram Launch, Authors, Powell and Maise

- Figure 5.4 Ascent Trajectory to GEO Using StarTram Launch, Authors, Powell and Maise

- Figure 5.5: Layout of Gen-1 Cargo Craft Geometry, Authors, Powell and Maise

- Figure 5.6 Japan Rail Maglev Passenger Vehicle with Mt. Fuji, Author, Japan Rail

- Figure 5.7 Drawing of Elevated Maglev Guideway Alongside US Interstate Highway, Author, Douglas Rike

- Figure 5.8 Gen-1 and Gen-2 Launch Systems, Author, Powell and Maise

- Figure 5.9 View of Gen-2 StarTram Exiting the Launch Tube, Author, John Rather

- Figure 5.10 Earthmap 1000x500 compac, http://en.wikipedia.org/wiki/File:Earthmapcompac.jpg, Author, jimhtofshaw tot ca, modification by Rodingold

- Figure 5.11 Space to Ground Microwave using Laser Pilot, http://en.wikipedia.org/wiki/file:space _to_ground_microwave_pilot_beam.png, Author NASA

- Figure 5.12 Suntower Space Solar Power Satellite, http://en.wikipedia.org/wiki/File:Suntower.jpg. Author, NASA

- Figure 5.13 Sandwich Space Solar Power Satelliote, http://en.wikipedia.org/wiki/file:solar_power_satellite_sandwich_or_abacus_co ncept.jpg. Author, NASA

- Figure 5.14 Space Solar Power Satellite Using Magnetically Inflated Cable Loop, Authors, Powell and Maise

- Figure 5.15 Space Solar Power Satellite Using Magnetically Inflated Cable Solar Concentrator, Authors, Powell and Maise

## Chapter 6

- Figure 6.1 US Electrical Generation Capacity, Author, Energy Information Adminstration (EIA) Electrical Power Annual 2007

- Figure 6.2 Average Capacity Factor by Source, Author, Energy Information Administration (EIA) Electric Power Annual 2007

- Figure 6.3A Google Earth Views of Black Hill near Boulder City, Nevada and Buffalo Mountain near Oak Ridge, Tennessee

- Figure 6.3B Buffalo Mountain near Oak Ridge, Tennessee

- Figure 6.4 Isometric View of Maglev Energy Storage Vehicles on Guideway, Author, Powell

- Figure 6.5 Cross Section View of Maglev Energy Storage Vehicle with Concrete Block on Guideway, Author, Powell

- Figure 6.6 Cross Section View of Maglev Energy Storage Vehicles without Concrete Block on Guideway, Author, Powell

- Figure 6.7 Cross Section View of Maglev Energy Storage Vehicle and Guideway on Concrete Pavement, Author, Powell

- Figure 6.8 Drawing of Guideway Beam and Panels, Author, Powell

- Figure 6.9 Drawing of MAPS Energy Storage Mode, Author, Powell

- Figure 6.10 Drawing of MAPS Energy Delivery Mode, Author, Powell

280

- Figure 6.11 Drawing of Layout of MAPS Energy Storage Facility, Author, Powell

- Figure 6.12 Potential Types of MAPS Locations, Author, Powell

- Figure 6.13 View of Japan Railways Superconductor Maglev Guideway and Vehicle System. Author, Japan Rail

- Figure 6.14 Artists Drawing of Maglev 2000 Vehicle on Monorail Guideway and vehicle System, Author, Japan Rail

- Figure 6.15 NbTi Superconductor Loop for Maglev 2000 Quadrupole, Authors Powell and Danby

- Figure 6.16 Superconducting Loop Enclosed in Stainless Steel Jacket, Authors, Powell and Danby

- Figure 6.17 Completed Guideway Panel with Figure-of-Eight Dipole and LSM Propulsion Loops, Authors, Powell and Danby

- Figure 6.18 Polymer Concrete Panel with Enclosed Aluminum Loop, Authors, Powell and Danby

- Figure 6.19 Photo of 72 foot long Monorail Guideway Beam Delivered from Construction Site in New Jersey to to Maglev 2000 Facility in Florida, Authors, Powell and Danby

## Chapter 7

- Figure 7.1 US Energy-Related Carbon Dioxide Emissions by Sector 2011, Author, US EIS, Monthly Energy Review, Tables 12.2-12.5 (May 2012)

- Figure 7.2 US Energy Flows: 2009, Author, US EIA Annual Energy Review, 2009

- Figure 7.3 Automobiles per 1000 persons as a function of GDP per Capita for Different Countries, Authors, Powell and Jordan

- Figure 7.4 Thomas Parker Electric Car, http://commons.wikimedia.org/wIki/File:Thomas_Parker_Electric_car, Author,Unknown

- Figure 7.5 Photo of Thomas Edison with Electric Car 1913, https: commons, Wikimedia.org/wiki/File: Edison_Electric_Car.jpg, Author, Smithsonian

- Figure 7.6 1960 Henny Kilowatt Electric Car, https://commons.wikimedia.org/wiki/file:kilowatt.jpg, Author, D. Roberson

- Figure 7.7 Tesla Roadster, https:commons.wikimedia.org/wiki/File:Roadster_Goodwood.jpg Author, Tesla Motors, Inc.

- Figure 7.8 NREL FT Diesel vs conventional Diesel photo https://commons.wikimedia.org/wiki/File:NREL_FT_diesel_vs_colnventional_diesel_photo.jpg, Author, DOE

- Figure 7.9 OSD Clean Fuel Initiative FT Diesel Emissions Presentation, http://wikipedia.org/wiki/File:OSD_Clean_Fuel_Initiative_FT_Diesel_Emission_pr esentation, Authors, Dr. Theodore.K.Bornu, Edward Sheridan, William E. Harnson

- Figure 7.10 Denali, Mt. McKinley, https: commons.wikimedia.org/wiki/File:Denali_Mt_McKinley.jpg, Author, National Park Service

- Figure 7.11 View of Mississippi River from Fire Point in Effigy Mounds National Monument, http://commons.org.wiki/file:Efme_view_from_fire_point, Author, National Park Service

- Figure 7.12 Windmill Farm, http:www.flcr.com/photos/57904403@No5/5532750017/?rb+1, Author, Unknown

- Figure 7.13 Statue of Liberty, http://commons.wikimedia.org/wiki/File:Detroit_photographic, Author, Unknown

## Chapter 8-No Figures

## Chapter 9

- Figure 9.1 Rembrandt-Belshazzar's Feast, https:commons.wikimedia.org/wiki/File:Rembrandt_Belshazzars_Feast_WGA191 23.jpg. Author, Rembrandt

- Figure 9.2 Michelangelo – Delphic Sybil, https:commons.wikimedia.org/wiki/File: Michelangelo_Delphic_Sybil.jpg. Author, Michelangelo

- Figure 9.3 Cassandra1, https:commons.wikimedia.org/wiki/File:Cassandra1.jpg. Author, Evelyn De Morgan

- Figure 9.4 Garden of Earthly Delights, https://commons.wikimedia.org/wiki/File:The_Garden_of_Earthly_Delights_High _Resolution.jpg. Author, Heteronomous Bosch

## Biographies of Author and Contributors

**James R. Powell, Ph.D.** is a Director of the MAGLEV 2000 of  Florida Corporation. Dr. Powell and his colleague, Dr. Gordon Danby are the recipients of the 2000 Benjamin Franklin Medal in Engineering for their invention of superconducting Maglev. The medal was awarded by The Franklin Institute "for their invention of a magnetically-levitated transport system using super conducting magnets and subsequent work in the field." The Franklin Institute awards medals annually in recognition of the recipients' genius and civic spirit and in memory of the Institute's namesake, Benjamin Franklin, who exhibited those same qualities. Some noted past recipients of the Franklin Institute medals include Alexander Graham Bell, Thomas Edison, Neils Bohr, Max Planck, Albert Einstein and Stephen Hawking.

He was a senior scientist at Brookhaven National Laboratory (BNL) from 1956 through 1996. His experiences have led to significant advances in the design and analysis of advanced reactor systems, cryogenic and super conducting power transmission, plasma physics, mine safety, fusion reactor technology, electronuclear (accelerator) breeder systems, transmutation of nuclear wastes, space nuclear thermal propulsion, electromagnetic hypervelocity guns, hydrogen and synthetic fuels, and transportation infrastructure.

He holds patents for the Particle Bed Reactor (PBR) for nuclear rocket propulsion, the use of aluminum structure in fusion reactors; blankets employing solid lithium ceramics and alloys for tritium breeding; and, demountable super conducting magnet systems and the Advanced Vitrification System (AVS) for high-level nuclear and toxic wastes. He and Dr. Danby are the holders of the first patent for superconducting Maglev in 1968, as well as many recent patents on their 2nd generation advanced maglev system.

Dr. Powell holds a Bachelor of Science in Chemical Engineering from the Carnegie Institute of Technology and a Doctor of Science in nuclear engineering earned in 1958 from the Massachusetts Institute of Technology. Dr. Powell has published almost 500 professional papers and reports. He is a member of the American Nuclear Society.

**Dr. Jesse Powell, Ph.D.** is Founder and President of Maglev Strategies, in which capacity he works to identify new markets and opportunities for maglev technologies. He coordinates between Maglev 2000, Inc. and third party companies in the scoping of new projects, and manages technology transfer issues. Currently, he is focused on maglev space launch, maglev energy storage, and maglev water transport as the areas most likely to attract funding in the United States.

Jesse Powell at Brookhaven National Lab on the Occasion of the 50th Anniversary of the invention of Superconducting Maglev by Gordon Danby and his father, James Powell
https://www.bnl.gov/video/index.php?v=514

From 2002 to 2013, Dr. Powell has worked in the field of Oceanography. He worked at Scripps Institution of Oceanography, where he studied the impact of ocean fronts and mesoscale ocean structures on plankton distributions. During this time, he also worked on a range of technology projects spanning from mission designs for the exploration of Mars and Europa, to the use of autonomous underwater vehicles to map ocean life, to machine vision systems for plankton identification and the automatic classification of fish eggs of important species for habitat mapping.

Dr. Powell holds a BS in Biology and BA in French Literature from University of California, San Diego, a MS in Molecular Biology from San Diego State University, and a PhD from Scripps Institution of Oceanography.

**James Jordan** is the founder and President of the Interstate Maglev Project and Executive Vice President of Maglev 2000. He is also the managing editor of The Fight for Maglev, Maglev America, and Silent Earth.

The energy crises of the 1970s focused the Navy career of James Jordan. The new era of scarce oil and rapid increases in oil prices dramatically introduced Commander Jordan to the military and economic security consequences of America's growing dependence on oil. Commander Jordan served as the first director of the Navy Energy R&D program office in the Pentagon. As director, he developed strategies and technologies aimed at sustaining military and national economic security in the new oil reality.

In 1979, Mr. Jordan retired from the Navy and became a senior policy advisor to the late Senator John C. Stennis, Chairman, Armed Services Committee and Defense Appropriations Committee. In this capacity, Mr. Jordan was a Senate staff leader in energy, transportation, environment, and agricultural policy.

In 1988, after leaving the U.S. Senate, Mr. Jordan founded several entrepreneurial ventures directed toward development of environmentally sustainable energy, and economic growth: efficient all-electric **mag**netic **lev**itation) transportation, nuclear waste isolation, advanced nuclear power generation and earth science data management.[1]

Education: MBA, Harvard Business School, Cambridge, MA; Distinguished Graduate, Industrial College of the Armed Forces at the National Defense University, Washington, DC; B.A., University of North Carolina, Chapel Hill, NC, Student Body President and Graduate of Senior High School, Greensboro, NC.

---

Consortium International Earth Science Information Network, (www.ciesin.org), now located at Columbia University. In 1992, Mr. Jordan introduced CIESIN to the U.N. Conference on Environmental Development (UNCED) in Rio.

# CURRICULUM VITAE

**Dr. James R. Powell**
**J&M Technologies P.O. Box 547**
**Shoreham, NY 11786**

EDUCATION

1953 - B.S., Carnegie Institute of Technology
1958 Sc.D., Chemical Engineering (Nuclear Engineering), MIT

CURRENT EMPLOYMENT

| | |
|---|---|
| 1997 - Present | President, DPMT Corporation and Director, Maglev 2000 of Florida Corporation. Development of Maglev transport system for Florida. President, Plus Ultra Technologies, Inc. Development of advanced space propulsion systems. Vice President and Chief Scientist, Radiation Isolation Consortium, LLC. Development of an advanced vitrification system for high level nuclear waste. |
| 1996-1993 | Head, Energy Systems Group, Division of Engineering Research and Applications, Department of Advanced Technology, Brookhaven National Laboratory. Work on new technologies for infrastructure, nuclear waste disposal, and neutron capture medical therapy. |
| 1993-1987 | Head, Reactor Systems Division, Department of Nuclear Energy, Brookhaven National Laboratory. Directed work on PBR SNTP (Space Nuclear Thermal Propulsion Program) and defense applications. |
| 1987-1974 | Head, Fusion Technology and Advanced Systems Group, Department of Nuclear Energy and Department of Applied Science, Brookhaven National Laboratory. Directed work on fusion/reactor technology and accelerator/ reactor hybrid systems. |
| 1980-1977 | Associate Chairman, Department of Nuclear Energy, Brookhaven National Laboratory. |

| 1974-1956 | Nuclear Engineer, Department of Applied Science and Department of Nuclear Engineering, Brookhaven National Laboratory. |
|---|---|

## AWARDS

Awarded Franklin Medal 2000 for Engineering, together with Dr. Gordon Danby, for their invention of Superconducting Maglev

## ORGANIZATIONS, COMMITTEES, MEMBERSHIPS

AIAA Highly Reusable Space Transportation (HRST) Study Maglev Task Force, U.S. Senate, Energy and Public Works Committee (Co-Chairman)

- National Academy of Science Study for Inertial Fusion
- National Academy of Science Study on Future Aircraft Carriers
- DOD Advisory Panel on Electro-magnetic Guns and Launchers American Institute of Astronautics

James R. Powell received his B.S. in chemical engineering in 1953 from the Carnegie Institute of Technology and his Sc.D. in nuclear engineering from the Massachusetts Institute of Technology in 1958. He joined Brookhaven National Laboratory (BNL) in 1956 where he was a Senior Nuclear Engineer and a tenured member of the BNL scientific staff until his retirement at the end of 1996.

While at BNL, in addition to his personal research, Dr. Powell directed R&D efforts in a variety of areas, including advanced fission reactors, fusion reactors, space and defense systems, and new technologies for infrastructure. During this period, he published 470 papers and reports, and received 16 Patents.

Since his retirement from Brookhaven National Laboratory at the end of 1996, Dr. Powell has continued to work and publish in the areas of Maglev transport, advanced space propulsion systems, and advanced vitrification systems for high level nuclear waste.

His research has been carried out in a wide range of fields, covering both nuclear and non-nuclear applications, including:

- Nuclear Fission Reactors
- Fusion Technology
- Space and Defense Systems
- Transportation and Infrastructure
- Environment and Safety
- Energy Technology (non-nuclear)
- Superconducting Systems
- Accelerator and Medical Applications
- Plasma Physics

In these various fields, Dr. Powell has concentrated on innovating, developing, and implementing new technological capabilities, His work has been interdisciplinary in nature, seeking out common threads where possible, so as to use knowledge gained in one area to provide new capabilities and knowledge in other areas. His principal accomplishments in these fields include:

In the field of Nuclear Fission Reactors, Dr., Powell is the inventor of the Particle Bed-Reactor, an ultra-compact, very high power density reactor system with applications for high performance nuclear rockets and space power, nuclear waste transmutation, and very high flux neutron beams for research. He also has carried out innovative advanced reactor systems, including the LMFR (Liquid Metal Fueled Reactor), chemo-nuclear reactors, and nuclear MHD reactors, involving high temperature chemistry, non-equilibrium plasmas, and ionization induced chemical reactions.

In the field of Fusion Technology, Dr. Powell originated the concept of using solid lithium compounds for fusion blankets, and carried out the first designs and work in this area. This concept, which is now the standard approach in fusion reactor design, has replaced the old approach of liquid lithium blankets.

Dr. Powell also originated the concept of designing fusion blankets for minimum radioactivity; using appropriate materials, he showed that long-lived radioactivity could be reduced by a factor of ~106 , as compared to previous designs. Designing for minimum radioactivity is now customary for fusion reactors.

Other fusion related inventions by Dr. Powell include blankets for production of hydrogen based synthetic fuels; NOEL, a self-sealing, no-leak blanket; TRAIL, a pellet rail gun plasma limiter system; DEALS, a demountable superconducting toroidal field coil system for tokamaks, and liquid wall blankets for inertial pellet fusion reactors.

Dr. Powell was a member of the National Academy of Science Panel that reviewed and provided guidance on the DOE program on Inertial Fusion. He also organized workshops and symposia on fusion blanket design, synthetic fuel production, and superconducting magnet safety and reliability.

In the field of Space and Defense Systems, Dr. Powell is the inventor of the high performance Particle Bed Nuclear Rocket, which formed the basis for the DOD Space Nuclear 'Thermal Propulsion (SMP) program. In the SNTP program, which was funded by SDI and the Air Force from 1983 to 1987, the BNL Reactor Systems Division, led by Dr. Powell, developed and analyzed the PBR design, together with thermal hydraulic testing and the development of advanced high temperature materials and nuclear fuel particles. When the $200 million SNTP program was terminated in 1993 due to the ending of the Cold War and its mission, BNL, together with its partner organizations, Grumman Aerospace, Babcock and Wilcox, Sandia National Laboratory, Allied Signal, and Hercules, had developed and tested the various PBR components, along with a nuclear critical assembly and initial nuclear tests of high power PBR fuel elements preparatory to ground testing the prototype PBR nuclear rocket engine.

Dr. Powell also developed innovative designs for light weight nuclear space electric power generating systems based on the PBR for a variety of civilian and defense applications. During the period from 1972 to 1985, he was a member of the DOD Advisory Panel on Electromagnetic Guns and Launchers, which helped to guide the development of EM guns by DOD. In conjunction with this effort, he originated innovative concepts for space based hypervelocity guns for strategic defense applications, including the invention and demonstration of a new type of hypervelocity gas gun based on the rapid heating of low molecular weight propellant by a particle bed thermal storage device. This concept, in addition to defense applications, has the potential to provide a much cheaper way to launch materials into orbit than is now possible using rockets.

Dr, Powell was a member of the National Academy of Science study on Future Aircraft Carriers, and has participated in a number of other defense related study efforts. During the period from 1956 to 1996, he held a DOE (formerly the AEC) "Q" clearance and worked on a wide range of classified subjects-

Since retiring from Brookhaven in 1996, Dr. Powell has continued to work in the field of advanced space propulsion systems. He is a president of Plus Ultra Technologies, Inc. which was awarded an STTR contract from NASA to carry out R&D on a compact ultra-lightweight nuclear thermal propulsion engine for planetary science missions, along with contracts from the NASA Institute of Advanced Concepts for the study of new approaches for the human exploration of Mars and robotic flyers in the atmosphere of Jupiter.

In the field of Transportation, Dr. Powell, together with Dr. Gordon Danby, are the inventors of the Superconducting Maglev (magnetic levitation) transportation systems and received the original patent in the field. Their inventions, including the inductive levitation and stabilization guideway, null flux geometry, and the Linear Synchronous Motor for vehicle propulsion, have been adopted throughout the world, and form the basis for the 300 mile Tokyo to Osaka maglev route now under construction in Japan. Drs. Powell and Danby were co-chairmen of Senator Moynihan's 1989 Maglev Task Force. Their report to the Senate Committee on Environment and Public Works helped provide the basis for authorization of the ISTEA program for the development of Maglev in the U.S.

Dr. Powell is President of the Danby-Powell Technology Corporation and a Director of the Maglev 2000 of Florida Corporation. He and Dr. Danby are directing the development of a 20 mile Maglev route that will link Port Canaveral, the Kennedy Space Center, and the Space Coast Regional Airport. The 20 mile route will demonstrate on advanced Maglev system based on low cost prefabricated guideways, with the ability to carry freight as well as passengers, and to electronically switch at high speeds to off-line stations. He and Dr. Danby were awarded the Franklin Medal 2000 for Engineering in April 2000, for their invention of superconducting Maglev. Previous awardees have included Nikolai Tesla, Charles Steinmetz and Albert Einstein.

In the field of Infrastructure, Dr. Powell, together with Dr. Morris Reich, organized the NICEST (National Infrastructure Center for Engineering Systems and Technology) industry/national laboratory/academic consortium for the development of new infrastructure technologies that could dramatically reduce the cost of infrastructure construction and maintenance, and improve its performance. Among their inventions are MAGI, a new method for magnetically imaging in three-dimension underground arrays of metal pipes and rebars to determine their size, location, and condition, and RAPI'OR, a light weight hypervelocity gas gun that fires high velocity projectiles to cut or rubbelize concrete structures and pavements.

In the field of Environment and Safety, Dr. Powell carried out pioneering work in demonstrating the use of high precision photographic methods to monitor the condition and safety of underground mine tunnels, as well as the initial studies on the safety and reliability of large superconducting magnet systems for fusion reactors. More recently, he has been involved in environmental and safety studies of high level nuclear waste problems. Together with Dr. Morris Reich, he is the inventor of new small module systems to directly vitrify concentrated high level waste inside closed final disposal containers. This invention potentially could greatly simplify the problem of disposing of high level waste and greatly reduce cost. He is Vice President and Chief Scientist of the Radiation Isolation Consortium, which is developing the Advanced Vitrification System (AVS) based on his and Dr. Reich's inventions.

In the field of non-nuclear Energy Technology, Dr. Powell, together with Dr. John Karkeck, carried out the first detailed studies of the potential benefits to the U.S. energy economy of large scale district heating systems. He also proposed new technologies for more efficient energy generation, involving the use of metal hydrides for gas compression in closed Brayton cycles, and removal of particulates in combustion fired power systems using pulsed lasers (the LASH Concept).

In the field of Superconducting Systems, Dr. Powell, in addition to the invention of superconducting maglev transport, carried out pioneering work in the areas of superconducting magnet systems for fusion reactors (the DEALS demountable magnet system for tokamaks and the superconducting AC power transmission project).

In the field of Accelerator Applications, Dr. Powell, together with Drs. Meyer Steinberg and Hiroshi Takahashi, carried out the first studies of the application of high current linear accelerators for the transmutation of long-lived nuclear waste. They, together with Dr. Pierre Grand, also carried out the first studies of the application of accelerators for the production of tritium, which became the basis for the present APT (Accelerator Production of Tritium) project in DOE.

In the area of Medical Applications of Accelerators, Dr. Powell has recently developed two new accelerator target concepts, NIFTI and DISCOS, for the generation of neutrons for use in Boron Neutron Cancer Therapy (BNCT). Detailed studies carried out with Dr. Hans Ludewig indicate that these new target designs enable a factor of 10 reduction in the accelerator current needed for practical BNCT treatment facilities. When developed, this would result in

BNCT facilities being much lower in cost, so that they could be widely implemented in hospitals.

In the field of Plasma Physics, Dr. Powell was the first person to experimentally produce ball lightning-like long lived luminosities in the laboratory, and to demonstrate that metastable very long-lived high energy electronic states in oxygen and nitrogen was the principal mechanism acting in natural ball lightning. Based on this experimental work, he, together with Professor David Finkelstein, developed a realistic, consistent model describing natural ball lightning. He also developed, with Professor Finkelstein, the piezo-electric theory to explain the natural lightning-like phenomena that sometime accompany earthquake. This theory takes into account the observed long range crystallographic ordering of quartz-containing rock strata by tectonic stresses to produce structures that exhibit piezo-electric constants that are a significant fraction of the value for pure quartz crystals.

Dr. Powell also carried out innovative experiments as nuclear induced metastable excitation in plasmas as a potential ioniztion mechanism for nuclear MHD (magnetohydrodynamics) power cycles. He also carried out extensive experimental work on liquid metal pulsed MHD generators for space electric power generation, including the investigation of Rayleigh-Taylor instabilities of conducting metal surfaces in applied magnetic fields.

Dr. Powell has organized three AAAS symposia related to space and maglev transportation topics, along with workshops on fusion technology, and has chaired numerous sessions at various technical meetings.

# List of Maglev Reports

1. Powell, J. and Danby, G., 1966. "High Speed Transport By Magnetically Suspended Trains", Paper 66WA/RR-5, ASME Winter Annual Meeting, New York, NY. Also, Powell, J. and Danby, G., 1967 "A 300 mph Magnetically Suspended Train, Mech Eng 89, p. 30-35

2. Powell, J. and Danby, G., 1969. "Electromagnetic Inductive Suspension and Stabilization System For A Ground Vehicle"; US Patent 3,470,828

3. Powell, J. and Danby, G., 1969, "Magnetically Suspended Trains: The Application of Superconductors to High Speed Transport", Cryogenics and Industrial Gases, 4 (10), p.19

4. Powell, J. and Danby, G., (3. 1970. "Dynamically Stable Cryogenic Magnetic. Suspensions for Vehicles in Very High Velocity Transport Systems". <u>Recent Advances in Engineering Science</u>, Gordon and Breach, Vol 5: p. 159-182

5. Powell, J: and Danby, G., 1971. "The Linear Synchronous Motor and High Speed Transport" Proc Intersociety Energy Conversion Eng. Conference, Boston, MA, p. 11 8-131

6. Powell, J. and Danby, G., 1971. "Magnetic Suspension For Levitated Tracked Vehicles" Cryogenics <u>11</u>: p. 192-204

7. Danby, G. and Powell, J., "The Central Role of Cryogenics in Magnetically Levitated High Speed Trains" Proc. Of XIII International Conference on Refrigeration, Washington, DC

8. 8, Powell, J. and Danby, G., 1971. "Cryogenic Suspension and Propulsion systems for 200 —2000 mph Ground Transport" Proc. Cryogenic Society of America Conf. on Applications of Cryogenic Technology, Vol 4, p. 299-332

9. Danby G. and Powell, J., 1972. "Integrated Systems for Magnetic Suspension and Propulsion of Vehicles" Proc. 1972. Applied Superconductivity Conf., Annapolis, p.120-126

10. Powell, J. and Danby, G., 1972. "Integrated Magnetic Suspension and Propulsion Systems" Proc. IEEE Meeting of Industrial Applications Society. Philadelphia, PA (10 pages)

11. Danby, G., Jackson, J., and Powell, J. 1974. "Force Calculations for Hybrid (Ferro — Null Flux) Low Drag Systems, IEEE Trans on Magnetic Mag <u>10</u>, p. 443.446

12. Danby, G. and Powell, .1., 1974. "Hybrid Superconducting Magnetic Suspensions for Very Efficient High Speed Ground Transport" Proc. 1974, Applied Superconductivity Conference

13. Danby, G. and Powell, J., 1988. "Design Approaches and Parameters for Magnetically Levitated Transport Systems" Proc. 2nd Annual Conference on Superconductivity and its Applications, Elsevier Science Publishing

14. Powell, J,. 1992. "Large Scale Implementation of Maglev in the United States" AAAS Symposium on Maglev Transport, Chicago, IL (10 pages)

15. Powell, J. and Danby, G., 1995. "Passenger and Freight Maglev for the US'-, Proc of the Future Transportation Technology Conf., Costa Mesa, CA. SAE Paper 95-1921

16. Powell, J., 1995. "The Application of Maglev Technology to Intermodal Transportation". Proc of National Aviation and Transportation Center, 4th Annual Symposium on Global Intermodalism and Economic Development, July 24-27, 1995

17. 17, Powell, J. and Danby, G., 1996. "Integrating Passenger and Freight Service: Maglev Technology Approach", Proc of 514 Annual Symposium on Intermodal Transportation. Bordeaux, France (41 pages)

18. Powell, J. and Danby, G., 1998. "Transport by Magnetic Levitation", Encyclopedia of Applied Physics, Vol 22, p 233-261

19. Powell, J. and Danby, G., 1989. Co-Chairman, Maglev Technology Advisory Committee for US Senate Committee on Environment and Public Works, "Benefits of Magnetically Levitated Transport for the United States", Volume 1, Executive Summary, (30 pages) and Volume 2, Technical Report (238 pages)

20. Powell, J. and Danby, G., 2000. "Magnetic Levitation: A New Mode of Transport for the 21th Century", Lecture Given at the Award of the 2000 Franklin Medal for Engineering to Powell and Danby by the Franklin Institute (50 Pages)

21. Powell, J. and Danby, G., 2002. "Maglev 2000 'Transportation Technology", Final Report Vol 1 (347 pages) and Vol 2, (452 pages) Federal Railroad Administration and Florida Department of Transportation (849 total pages)

22. "Florida Maglev Deployment Program Kennedy Space Center Circulator National Demonstration Project", Tilden, Lobnitz & Cooper, 2002, (238 Pages)

23. Powell, J. and Danby, G., 2005. "Final Report, Federal Transit Administration, Maglev 2000 Project FL-26-7023", Maglev 2000 of Florida Corporation (100 pages)

24. Powell, J. "Maglev Presentation", ETA Low Speed Urban Maglev Workshop. W.Kulyk, Director, Office of Mobility Innovation, September 8.9, 2005, Washington, DC (30 pages)

25. Powell, J. and Danby, G. "Integration of Maglev Guideways with Railroad Track: The MERRI System", Report DPMT-1, October 5, 1996 (70 Pages)

26. Powell, J. and Danby, G. "Maglev Vehicles", IEEE Potentials, p.7-12, October/November, 1996

27. Powell, J. and Danby, G., "The Development of Maglev-Yamanashi and Beyond", Invited Talk at Dedication of Japan Railways Yamanishi Maglev Test Line, April 4, 1997) (23 Pages)/

28. Powell, J. and Danby, G., "The M-2000 Maglev System for the United States", Presentation to the FRA Maglev Advisory Committee, March 24, 1997 (28 Pages)

29. Powell, J. and Danby, G., "Maglev Technologies for Combined Freight and Passenger Movement — Application to Industrialized and Rapidly Industrializing Nations", 6th International Symposium in Intermodal Transportation, Mexico City June 18-20, 1997 (29 Pages)

30. Powell, J, and Danby, G., and Bragden, C., "Developing A High Speed Maglev Land Bridge in Central America" 66 International Symposium on Intermodal Transportation, Mexico City, June 18-20 1997 (28 pages)

31. Powell, J. and Danby, G., "The Water Train: Long Distance Transport of Water by Maglev", Report DPMT-3, December 13, 1997 (34 pages)

32. Powell, J., "Electrical Power Storage Using Maglev — The Maglev Power Storage System", Report DPMT-14, November 1, 2000 (126 Pages)

33. Powell. J. and Danby, G., "Low Speed Application of Superconducting Maglev", presentation to Transportation Research Board, January 14, 1999, Washington, DC (30 pages)

34. Powell, J., "Cost Projections for the M-2000 Maglev System" Report M-2000/002, presentation to the FRA Workshop, August 24, 1999, Washington, DC (30 pages)

35. Danby, G., "The M-2000 Maglev System" Report 2000/001, Presentation to the FRA Workshop, August 24, 1999, Washington, DC (26 Pages).

36. Danby G., and Powell, J., "Progress in Design and Testing of Maglev 2000 Technology", Report No. M-2001, Presentation to Transportation Research Board 2000 Annual Meeting, Washington, DC, January 9-13, 2000 (29 pages1

37. Powell, and Danby, G., "Maglev: An Evolutionary Technology for the Transport of Freight, People, and Resources", Presented at International Congress on the Implementation Follow Up of Habitat Agenda in Islamic Cities, Tunis, Tunisia, March 24-26, 2000 (46 pages)

38. Powell, J. and Danby, G., Morena, J, Wagner, T, and Smith, C., "The Maglev 2000 Urban Transit Systems", July 31, 2002 (15 Pages).

39. Powell, James, and Danby, Gordon, "The 2nd Generation Maglev 2000 Transport System: Design, Technology, Status, and Future Applications, Maglev 2000 of Florida, Sept. 2006 (166 pages)

40. Jordan, James, and Powell, James " Maglev Transport – A Necessity in the Age of No Oil", presented Capital Science 2008, Arlington, VA., National Science Foundation, March 29-30, 2008 (40 pages)

41. Ibid, PowerPoint presentation, March 29-30, 2008 (25 pages)

42. Powell, James and Gordon Danby, "Energy Efficiency and Economics of Maglev Transport", Presented at 2008 Advanced Energy Conference, Stony Brook University, NY, November 19-20, 2008 (22 pages)

43. Ibid, PowerPoint presentation, November 19-20, 2008 (18 pages)

44. Powell, James, et al, "A National Maglev Network for the US – Design and Capabilities", Presented at Maglev 2008, 20th International Conference on Magnetically Levitated Systems and Linear Drives. San Diego, California, December 15-18, 2008 (9 pages)

45. Danby, Gordon, et al, "Fabrication and Testing of Full-Scale Components For the 2nd Generation Maglev 2000 System, 20th International Conference on Magnetically Levitated Systems and Linear Drives, San Diego, California, December 15-18, 2008 (10 pages)

46. Powell, James, et al, "Adaptation of the LIRR System to Maglev for Faster, More Convenient, and Lower Cost Service", Presented at 2nd Advanced Energy Conference, Stony Brook, Long Island, New York, November 18-19, 2009 (18 pages)

47. Ibid, PowerPoint Presentation, Nov 18-19, 2009 (29 pages)

48. Griffis, F.H. (Bud), et al, "How Maglev Can Enable Stewart Airport to become the 4th Major Airport for the NYC Region". Presented at the 2nd Advanced Energy Conference, Stony Brook, Long Island, New York, November 18-19, 2009 (21 Pages)

49. Ibid, PowerPoint Presentation, Nov 18-19, 2009 (23 Pages)

50. Powell, James, et al, "The West Coast Maglev Network Transport for the 21st Century, Maglev 2000, January 20, 2010 (39 pages)

51. Powell, James, et al, "The New York Maglev Network – A New Transport System for the 21st Century, Maglev 2000, January, 2011 (12 pages)

52. Powell, James, et al, "Maglev Energy Storage and the Grid", Presented at the 2010 Advanced Energy Conference, New York, NY, November 8-9, 2010 (11 pages)

53. Griffis, F.H, (Bud), et al, "Feasibility Study for New Danby Powell Maglev System & Preliminary Route Study (New York City to Stewart International Airport), Preliminary Guideway Plans and Updated Cost Estimates, Polytechnic Institute of NY, January 2011 (144 pages)

54. Powell, James, et al, "The Maglev America Project: A 29,000 National Maglev Network for the United States", Presented at Maglev 2011, 21st International Conference on Magnetically Levitated Systems and Linear Drives, Daejeon, Korea, Oct 10-13, 2011 (15 pages)

55. Ibid, PowerPoint Presentation, October 10-13, 2011 (17 pages).

56. Griffis, F.H.(Bud), et al, "Adaptation of Existing Railroad Trackage for Levitated Maglev Vehicles", presented at Maglev 2011, 21st International Conference on Magnetically Levitated Systems and Linear Drives, Daejeon, Korea, Oct. 10-13, 2011 (6 pages)

57. Powell, James, et al, "Large Scale Storage of Electrical Energy Using Maglev", presented at Maglev 2011, 21st International Conference on Magnetically Levitated Systems and Linear Drives, Daejeon, Korea, Oct 10-13, 2011 (19 pages)

58. Maglev 2000 Proprietary Reports

59. Powell, J., "Geometry and Magnetic Forces on the Electromagnetic Loops for the M-2000 Narrow Beam Guideway", Memo M-2000-JP-3-01-6/14/96, June 14, 1996 (61 pages).

60. Powell, J., "Fringe Fields From SC Magnets", Memo M-2000-JP-3-02-7/26/96, July 26, 1996 (13 pages)

61. Powell, J., "Computer Analysis of Magnetic Field Distributions From M-2000 SC Quadrupole Arrays: Nature and Requirements for Phase A Studies", Memo M-2000-JP-03-8/03/96 (32 pages).

62. Powell, J., "Refrigeration Systems for High Temperature Superconducting Vehicle Magnets", Memo M-2000-JP-4-01-8/17/96, August 17, 1996 (6 pages).

63. Powell, J., and Maise, G, "Vehicle Design and Mass Budget", Memo M-2000-JP/GM-01-1-8/24/96 August 24, 1996 (14 pages)

64. Maise, G., "Potential Damage to Sides of Vehicles from Windborne Debris", Memo M-2000=GM-1-01-09/26/96 September 26, 1996 (4 pages)

65. Maise, G., "Aerodynamic Lift to Augment Magnetic Levitation", Memo M-2000-GM-1-03-12/11/96 December 11, 1996 (5 pages)

66. Powell, J., "Optimization of Placement of LSM Winding", Memo M-2000-JP-3-06-12/31/96, December 31, 1996 (5 pages).

67. Maise, G., "Vehicle Drag Coefficient", Memo M-2000-GM-1-02-10/31/96 October 31, 1996 (9 pages)

68. Maise, G., "Design of M-2000 Vehicle", Memo M-2000-GM-1-04-1/21/97, January 31, 1997 (6 pages).

69. Powell, J., "Vehicle Speeds and Accelerations on Proposed Brevard County Route", Memo M-2000-JP-01-02-1/10/97, January 10, 1997 (11 pages)

70. Maise, G., "Tilting of M-2000 Guideway and/or Vehicles to Eliminate Lateral Forces on Passengers" Memo M-2000-GM-1-05-07/11/97, July 11,1997 (6 pages)

71. M-2000 Program Review; Presentation to C. Smith, Florida Department of Transportation, August 1, 1947 (33 pages)

72. Maise, G., "Aerodynamic Loads on the Guideway Structure Due to Hurricane — Level Wind Forces", Memo M-2000 — GM-1-07/12/31/97, December 31, 1997 (10 pages)

73. Powell, J., "Design of Guideway Loops and Panels, Memo M-2000-JP-3-04-11/15/96, November 15, 1997 (24 pages)

74. Powell, J., "Reduction of Fringe Fields in the Passenger Cabin by Use of Asymmetric Quadrupoles", Report DPMT-6 December 15, 1997 (7 pages)

75. Powell, J., "The NOVI Ride Control System: A Method for Eliminating Vibration and Maximizing Ride Comfort in Maglev Transportation Systems", Report DPMT-4, December 15, 1997 (23 pages)

76. Powell, J., "On Surface Maglev Guideways", Report DPMT-2, September 15, 1997 (11 pages)

77. Powell, J., "Magnetic Anchoring of Maglev Guideway Beams", Report DPMT-5, December 15, 1997 (22 pages).

78. Powell, J., "Morena, G., Powell, J., and Danby, G., "Transport of Bulk Cargo for Above and Underground Mining by Low-Cost, High-Speed Magnetically Levitated Vehicles", Report M-205, National Mining Associates 21" Annual Transportation and Distribution Seminar, January, 1998 (15 pages)

79. Powell, J., ed., "Cost Projections for the M-2000 Maglev System", Report DPMT-20, May 15, 1999 (192 pages).

80. Powell, J., "The Matrushka Magnet – A Low Cost Ultra-Low Refrigeration Load Magnet System", Report DPMT-9, April 1998 (113 Pages)

81. Powell, J., "Non-conventional Methods for Large Scale Manufacture of Maglev Guideway Loops", Report DPMT-7, February 15, 1999 (31 pages)

82. Powell, J., "Analysis of Eddy Current Heating in Guideway Conductors and Determination of Allowable Limits on Conductor Size", Report DPMT-8, February 15, 1999 (37 pages)

83. Lazareth, O., Skaritka, J., and Powell, J. -- "Comparison of Analytical Calculations of Magnetic Forces for the M-2000 Maglev Systems with Experimental Measurements", Report DPMT-11, August 11, 1999 (35 pages)

84. Powell, J., "Power Transmission and Distribution Architecture for the M-2000 Maglev System" Report DPMT-12, December 15, 1999 (76 pages)

85. Powell, J., "Levitation, Propulsion, Power, and Braking Systems for Maglev 2000 Vehicles", Report DPMT-22, May 18, 2000 (42 pages)

86. Powell, J., "Communications, Control, and Safety for the M-2000 Maglev System', Report DPMT-23, May 18, 2000 (13 pages).

87. Maise, G., "Design of the M-2000 Maglev Passenger Vehicle, Report DPMT-21, May 30, 2000 (27 pages)

88. Powell, J., "The IRT Levitation Demonstration", Report DPMT-26, May 2001 (20 Pages)

89. Skaritka, J., "Superconducting Magnet Design', Report DPMT-24, May 2001 (23 Pages)

90. Harmer, E, Danby, G., Lazareth, O., and Powell, J., "Experimental Measurements of the Magnetic Forces Between the Maglev 2000 Superconducting Quadrupole and Powered Guideway Loops, with Comparison to Values Predicted Using the Maglev 2000 Computer Code", October 15, 2002 (57 pages)

91. Lazareth, O. and Powell, J., "Computer Analyses of Magnetic Forces Between the M-2000 Vehicle and Powered Guideway Loops", Report DPMT-25 (59 pages)

92. Lazareth, O. and Powell, J. "Planar Guideway Performance of Compact Urban M-2000 Revenue Vehicle Part L Levitation and Stability Performance on Non-Powered Guideway", Report M-2000 PG-2-1 (73 pages)

93. Lazareth, O. and Powell, J., "Planar Guideway Performance of Urban/Suburban M-2000 Maglev Revenue Vehicle, Part 1: Levitation and Stability Performance on Non-Powered Guideway" Report M2000 PG-1-1 (172 pages)

94. Lazareth, O. and Powell, J., "Propulsion and Power Performance of Maglev 2000 Vehicles on Planar Guideway", Report M-2000 PG-3-1, June 1, 2003 (114 pages)

# Detailed Papers and Reports.

- StarTram (Maglev Launch)
- MITEE (Nuclear Thermal Propulsion Engine)
- SUSEE (Nuclear Space Power Reactor)
- ALPH (Nuclear Robotic Probe and Factory)
- MIC (Magnetically Inflated Cable Space Structures)

# Detailed Papers and Reports on StarTram

1. StarTram: A New Concept for Very Low Cost Earth-to-Orbit Transport Using Ultra High Velocity Magnetic Launch, James Powell, George Maise, and John Paniagua, Paper IAF-01-S.6.04, 52nd International Astronautical Congress, 1-5 October 2001, Toulouse, France (36 pages).

2. Powell, J., and Maise, G., "Space Tram" US Patent No. 6,311,926B1, November 6, 2001.

3. StarTram: A New Approach for Low-Cost Earth-to-Orbit Transport, James Powell, George Maise, and John Paniagua, 2002 IEEE Space Conference, March 2002, Big Sky, Montana (17 pages).

4. StarTram C—A Maglev System for Ultra Low Cost Launch of Cargo to LEO, GEO, and the Moon, James Powell, George Maise, and John Paniagua, Paper IAC-03-IAA13.1.04, 54th International Astronautical Congress, October 2003, Bremen, Germany (18 pages).

5.  StarTram: The Key to a Robust, Low Cost Earth/Lunar Transport System, James Powell, George Maise, and John Paniagua, International Lunar Conference 2003, November 16-22, 2003, Hawaii (23 pages).

6.  StarTram: Ultra Low Cost Launch for Large Space Architectures, James Powell, George Maise, and John Paniagua, STAIF 2004 Conference, February 2004, Albuquerque, New Mexico (12 pages).

7.  StarTram: The Key to Low Cost Lunar Bases and Human Exploration of Space, James Powell, George Maise, and John Paniagua, AIAA Space 2004, September 28-30, 2004, San Diego, California (12 pages).

8.  StarTram: An Ultra Low Cost Launch System for Large Scale Exploration and Commercialization of Space, James Powell, George Maise, and John Paniagua, Paper IAC-04-V.05.07, 55th International Astronautical Congress, October 2-8, 2004, Vancouver, Canada (17 pages).

9.  Ibid, StarTram viewgraphs presented at the 55th IAC meeting, Vancouver, Canada, (26 pages).

10. StarTram: An Ultra Low Cost Launch System to Enable Large Scale Exploration of the Solar System, James Powell, George Maise, and John Paniagua, STAIF 2006 Conference, February 12-15, 2006, Albuquerque, New Mexico (12 pages).

11. StarTram: An International Facility to Magnetically Launch Payloads at Ultra Low Unit Cost, George Maise, James Powell, John Paniagua, and James Jordan, Paper IAC-06-D3.2.7, 57th International Astronautical Congress, October 2-5, 2006, Valencia, Spain (14 pages).

12. Ibid: StarTram viewgraphs presented at the 57th IAC Meeting, Valencia, Spain (25 pages).

13. StarTram: The Maglev Launch Path to Very Low Cost, Very High Volume Launch to Space; presented at the 14th International EML Symposium, Victoria, Canada, June 10-13, 2008.

14. The Gen-1 Maglev Launch System for Ultra Low Cost Access to Space; James Powell, George Maise, and John Paniagua; presented at the 59th International Astronautical Congress (IAC), Glasgow, Scotland, September 29—October 3, 2008 (10 pages).

15. Ibid: Viewgraphs presented at the 59th IAC Meeting, Glasgow, Scotland (21 pages).

16. Maglev Launch—An Ultra Low Cost Way to Deploy Space Solar Power Systems; presented at the From the Sun to the Earth International Conference on solar Energy from Space, Ontario Science Center, Toronto, Canada, September 8-10, 2009 (17 pages).

17. Ibid: Viewgraphs presented at the From the Sun to the Earth Conference, Toronto, Canada (33 pages).

18. Maglev Launch: Ultra Low Cost, Ultra/High Volume Access to Space for Cargo and Humans; James Powell, George Maise, and John Rather, presented at SPESIF-2010—Space, Propulsion, and Energy Sciences International Forum, February 23-26, 2010, Johns Hopkins Applied Physics Laboratory, Baltimore, Maryland (15 pages).

19. Ibid: viewgraphs presented at SPESIF-2010 meeting, Baltimore, Maryland (33 pages).

20. A Development and Test Program for the Generation-1 Maglev Launch System, James Powell, George Maise, and John Rather, presented at SPESIF-2011 Meeting, Baltimore Maryland (16 pages).

21. Ibid: Viewgraphs presented at SPESIF-2011 Meeting, Baltimore, Maryland.


# Detailed Papers and Reports on MITEE

1. MITEE: An Ultra Lightweight Nuclear Engine for New and Unique Planetary Science and Exploration Missions. James Powell, John Paniagua, George Maise, Hans Ludewig and Michael Todosow, Paper IAF-98-R.1.01, 49th International Astronautical Congress, Sept. 28—Oct. 2, 1998, Melbourne, Australia [27 pages].

2. Europa Sample Return Mission Utilizing MITEE Technologies. John Paniagua, James Powell, George Maise, Hans Ludewig, Michael Todosow, Paper IAF-98-Q.2.03, 49th International Astronautical Congress, Sept. 28—Oct. 2, 1998, Melbourne, Australia [15 pages].

3. Exploration of Jovian Atmosphere Using Nuclear Ramjet Flyer. George Maise, James Powell, John Paniagua, Hans Ludewig, Michael Todosow, Paper IAF-98-S.6.08, 49th International Astronautical Congress, Sept. 28—Oct. 2, 1998, Melbourne, Australia [11 pages].

4. High Performance Nuclear Thermal Propulsion System for Near Term Exploration Missions to 100 AU and Beyond. James Powell, John Paniagua, George Maise, Hans Ludewig, and Michael Todosow, Acta Astronautica, <u>44</u> No. 2-4, pp 159-166, Jan—Feb. 1999 [8 pages].

5. The Liquid Annular Reactor System (LARS) for Deep Space Exploration. George Maise, John Paniagua, James Powell, Hans Ludewig, and Michael Todosow, 2nd IAA Symposium on Realistic Near-Term Advanced Scientific Space Missions. June 29—July 1, 1998, Aosta, Italy; also Acta Astronautica, <u>44</u> No. 2-4, pp 167-174, Jan—Feb 1999 [13 pages].

6. New Approaches for the Exploration and Colonization of the Solar System: Road Map for the Next 30 Years in Space. James Powell, George Maise, and John Paniagua, Report PUR-7, Nov. 10, 1998 [18 pages].

7. The MITEE Family of Compact, Ultra Lightweight Nuclear Thermal Propulsion Engines for Planetary Space Exploration. James Powell, George Maise, and John Paniagua, Paper IAF 99-5.6.03, 50th International Astronautical Congress, October 4-8, 1999, Amsterdam, the Netherlands [28 pages].

8.  SunBurn: A Concept Enabling Ultra High Spacecraft Velocities for Extra Solar System Exploration. George Maise, James Powell, and John Paniagua, Paper IAA-99-IAA.4.1.07, 50th International Astronautical Congress, October 4-8, 1999, Amsterdam, the Netherlands [18 pages].

9.  A Cost Effective Space Infrastructure for Retrieval of Helium-3 from Uranus for Earth-Based Fusion Power Systems Utilizing the MITEE Nuclear Propulsion System. John Paniagua, James Powell, and George Maise, Paper IAA-99-R.3.10, 50th International Astronautical Congress, October 4-8, 1999, Amsterdam, the Netherlands [17 pages].

10. Compact, Ultra Lightweight Nuclear Thermal Propulsion Engines for Planetary Science Missions. James Powell, George Maise, John Paniagua, Hans Ludewig, and Michael Todosow, 10th Annual NASA/JPL/MFSC/AIAA Advanced Propulsion Research Workshop, Huntsville, Alabama, April 6-8, 1999 [25 pages].

11. Phase 1 Final Report: Lightweight High Specific Impulse (1000 sec) Space Propulsion Systems. James Powell, George Maise, John Paniagua, Jon Longtin, John Metzger, and Hui Zhang, PUR-12, October 1999 [221 pages].

12. Phase 1 NIAC Final Report: Exploration of Jovian Atmosphere Using Nuclear Ramjet Flyer. George Maise, James Powell, John Paniagua, and Robert Lecat, PUR-16, Nov. 30, 2000 [14 pages].

13. MITEE-B: A Compact Lightweight Bi-Modal Nuclear Engine to Deliver Both High Propulsive Thrust and High Electric Power. James Powell, George Maise, John Paniagua, and Stan Borowski, Paper IAF-01-S.6.05, 52nd International Astronautical Congress, Oct. 1-5, 2001, Toulouse, France [24 pages].

14. Europa One—A Manned Base for Exploration of the Outer Solar System and Near Interstellar Space. John Paniagua, James Powell, and George Maise, 52nd International Astronautical Congress, October 1-5, 2001, Toulouse, France [26 pages].

15  Phase 1 NIAC Final Report: Europa Sample Return Mission Utilizing High Specific Impulse Propulsion Refueled with Indigenous Resources. John Paniagua, James Powell and George Maise, November 30, 2001, Report PUR-21 [114 pages].

16. Compact MITEE-B: Bi-Modal Nuclear Engine for Unique New Planetary Science Missions. James Powell, George Maise, John Paniagua, and Stanley Borowski, AIAA 2002-3652, AIAA/ASME/SAE/ASEE Joint Propulsion Conference and Exhibit, July 2002, Indianapolis, Indiana [22 pages].

17. Phase 2 NIAC Interim Report: Exploration of Jovian Atmosphere Using Nuclear ramjet Flyer. George Maise, et al, PUR-26, Jan. 31, 2002 [125 pages].

18. Europa Sample Return Mission Utilizing High Specific Impulse Refueled with Indigenous Resources. John Paniagua, James Powell, and George Maise, Paper IAC-02-Q.2.05, 53rd International Astronautical Congress, October 10-19, 2002, Houston, Texas [14 pages].

19. Missions Possible: How Humanity Can Really Explore the Solar System Using Nuclear Propulsion, Report PUR-27, James Powell, George Maise, and John Paniagua, April 15, 2002 [22 pages].

20. Bi-Modal MITEE Engine for Nuclear Thermal/Nuclear Electric Propulsion. James Powell, George Maise, and John Paniagua, Advanced Space Propulsion Workshop, June 4-6, Pasadena, California [31 pages].

21. Exploration of Jovian Atmosphere Using Nuclear Ramjet Flyer. James Powell, George Maise, and John Paniagua, Advanced Space Propulsion Workshop, June 4-6, Pasadena, California [31 pages].

22. NEMO: Exploration of Europa's Subsurface Ocean and Return of Samples to Earth Using Nuclear Propulsion. James Paniagua, James Powell, and George Maise, Advanced Space Propulsion Workshop, Huntsville, Alabama, April 15-17, 2003 [35 pages].

23. MITEE-B: A Compact Ultra Lightweight Bi-Modal Nuclear Propulsion Engine for Robotic Planetary Science Missions. James Powell, George Maise, John Paniagua, and Stanley Borowski, STAIF 2003 Meeting, February 2003, Albuquerque, New Mexico [9 pages].

24. Pluto Orbiter/Lander/Sample Return Missions Using the MITEE Nuclear Engine. James Powell, George Maise and John Paniagua, 2003 IEEE Aerospace Conference, Big Sky, Montana, March 2003 [24 pages].

25. Exploration of Jovian Atmosphere Using Nuclear Ramjet Flyer. George Maise, et al., Phase II Final Report, March 1, 2003, NIAC Phase II Grant 07600-061 [163 pages].

26. HIP: A Hybrid NTP/NEP Propulsion System for Ultra Fast Robotic Orbiter/Lander Missions to the Outer Solar System. James Powell, George Maise, and John Paniagua, 54th International Astronautical Congress, October 2003, Bremen, Germany [17 pages].

27. MITEE and SUSEE: Compact Ultra Lightweight Nuclear Power Systems for Robotic and Human Exploration Mission. James Powell, George Maise, and John Paniagua, Paper IAC-04-IAA R.4/S.7-04, 55th International Astronautical Congress, Vancouver, Canada, October 2-8, 2004 [15 pages].

28. NEMO: A Mission to Explore and Return Samples from Europa's Oceans. James Powell, John Paniagua, and George Maise, STAIF 2004 Conference, February 2004, Albuquerque, New Mexico [7 pages].

29. Nuclear Propulsion and Power Systems for Near Term Exploration of the Solar System. James Powell, George Maise, and John Paniagua, AIAA 1st Space Exploration Conference, Jan 30—Feb. 1, 2005, Orlando, Florida [17 pages].

30. NEMO: A Mission to Search for and Return to Earth Possible Life Forms on Europa, Jesse Powell, James Powell, George Maise, and John Paniagua, Acta Astronautica, 57 pp 579-593, 2005 [15 pages].

31. Mini-MITEE: Ultra Small, Ultra Light NTP Engines for Robotic Science and Manned Exploration Missions. James Powell, George Maise, and John Paniagua, STAIF 2006 Conference, February 12-16, 2006, Albuquerque, New Mexico [10 pages].

32. MITEE: A Compact Near Term NTP Engine for New and Unique Robotic and Manned Exploration Missions. James Powell, George Maise, and John Paniagua, American Nuclear Space Conference, June 5-9, 2005, San Diego, California [9 pages].

33. The MITEE Hopper: A Compact NTP Spacecraft to Explore Multiple Surface Sites Using In-Situ Propellants. James Powell, George Maise, and John Paniagua, Paper IAC-06-D2.8/C3.5/C4.7/D3.5.06, 57th International Astronautical Congress, October 2-6, 2006, Valencia, Spain [12 pages].

34. A New Mission for the International Space Station (ISS) Enabled by Nuclear Thermal Propulsion—Cyclic Transport of Personnel and Supplies Between the Earth and the Moon, John Paniagua, James Powell, and George Maise, STAIF 2008 Conference, February 2008, Albequerque, New Mexico [9 pages].

35. Design and Development of the MITEE-B Bi-modal Nuclear Propulsion Engine, John Paniagua, James R. Powell, and George Maise.

36. Application of the MITEE Nuclear Ramjet for Ultra Long Range Flyer Missions in the Atmospheres of Jupiter and Other Giant Planets, George Maise, James Powell, John Paniagua, Edward Kush, Pasquale Sforza, and Hans Ludewig.


## Detailed Papers and Reports on SUSEE

1. SUSEE: Ultra Light Nuclear Space Power Using the Steam Cycle, James Powell, George Maise, and John Paniagua, IEEE 2002 Space Conference, Big Sky, Montana, 2002 [17 pages].

2. SASSE: A Lightweight, High Efficiency Solar Thermal Steam Cycle For Satellites, James Powell, George Maise, and John Paniagua, Paper IAC-03-R.2.07, 54th International Congress, Bremen, Germany, 2003 [16 pages].

3. SUSEE—An Ultra Lightweight Nuclear Electric Propulsion System Based on Existing Water and Steam Cycle Technology, James Powell, George Maise, and John Paniagua, Advanced Space Propulsion Workshop, Huntsville, Alabama, April 15-17, 2003 [31 pages].

4. Compact Ultra Light Nuclear Electric Power Systems for Future Moon Bases and Colonies, James Powell, George Maise, and John Paniagua, International Lunar Conference 2003, Hawaii, November 16-22, 2003 [13 pages].

5. MITEE and SUSEE: Compact Ultra Lightweight Nuclear Power Systems for Robotic and Human Exploration Missions, James Powell, George Maise, and John Paniagua, Paper IAC-04-IAA-R.4/S.7-04, 55th International Astronautical Congress, October 2-8, 2004, Vancouver, Canada [15 pages].

6. Ibid: Viewgraph presentation [29 pages].

7. SUSEE: An Ultra Light Space Nuclear Power System Based on Conventional Water Reactor Technology, George Maise, James Powell, and John Paniagua, American Nuclear Society, June 5-9, 2005, San Diego, California [9 pages].

8. SUSEE: A Compact, Lightweight Space Nuclear Power System Using Present Water Reactor Technology, George Maise, James Powell, and John Paniagua, STAIF 2006 Conference, February 12-16, 2006, Albuquerque, New Mexico [12 pages].

## Detailed Papers and Reports on ALPH

1. ALPH—A Robotic Precursor to Produce Large Amounts of Supplies for Manned Outposts on Mars, James Powell, John Paniagua, and George Maise. Presented at 49th International Astronautical Congress, Melbourne, Australia, September 28—October 2, 1998, Paper IAF-98Q.3.08 [38 pages].

2. MICE: A Compact, Light Near Term Mobile Robot for Exploration of the Martian Polar Ice Cap, James Powell, George Maise, John Paniagua, Hans Ludewig, and Michael Todosow, Presented at 50th International Astronautical Congress, Amsterdam, the Netherlands, October 4-8, 1999, Paper IAF 99-Q.3.08 [18 pages].

3. Development of Self-Sustaining Mars Colonies Utilizing North Polar Cap and the Martian Atmosphere, James Powell, George Maise, John Paniagua, and Jesse Powell, Final Report, NIAC Research Grant 07600-053, November 20, 2000 [184 pages].

4. Self-Sustaining Mars Colonies Utilizing the North Polar Cap and Martian Atmosphere, James Powell, George Maise, and John Paniagua, Presented at 51st International Astronautical Congress, Rio de Janeiro, Brazil, Oct. 2-6, 2000, also published in Acta Astronautica, 48, No. 5-12, pp. 737-765 (2001) [27 pages].

5. The Mars Hopper—A Mobile Lightweight Probe to Explore and Return Samples from Many Widely Separated Locations on Mars, James Powell, George Maise, and John Paniagua, Presented at 52nd International Astronautical Congress, Toulouse, France, Oct. 1-5, 2001, Paper IAA-01-IAA13.3.08 [26 pages].

6. Fast Track Route to Mars Colony Using Nuclear Propulsion and Power, James Powell, George Maise, and John Paniagua, Presented at 40th Aerospace Sciences Meeting and Exhibit, Reno, Nevada, Jan. 14-17, 2002, Paper AIAA 2002-0996 [32 pages].

7. CADMUS—A Robotic Mars Factory Returning Supplies to Earth Orbit, James Powell, George Maise, and John Paniagua, Presented at 2003 IEEE Aerospace Conference, Big Sky, Montana, March 2003 [18 pages].

8. Xanadu: A Polar Base for Manufacturing Supplies on Mars, James Powell, George Maise, and John Paniagua, Presented at STAIF 2004 Conference, Albuquerque, New Mexico, February 2004 [9 pages].

9.  Multi-MICE: A Network of Interacting Nuclear Cryoprobes to Explore Ice Sheets on Mars and Europa, Jesse Powell, James Powell, George Maise, and John Paniagua, Presented at Space 2005 Conference, Long Beach, California, Sept. 2005 [14 pages].

10. MERIT: A New Approach for a Large Scale Space Infrastructure Based on Mars, James Powell, George Maise, and John Paniagua, Presented at 2005 STAIF Conference, February 13-17, 2005, Albuquerque, New Mexico [10 pages].

11. ALPH: A Low Risk, Cost Effective Approach for Establishing Manned Bases and Colonies on Mars, James Powell, George Maise, John Paniagua, and Jesse Powell, Presented at AIAA Space 2005 Conference, Long Beach, California, August 30—Sept. 1, 2005 [17 pages].

12. ALPH: A compact Robotic Nuclear Powered Factory to Build and Supply Bases on Mars Prior to Manned Landing, James Powell, George Maise, and John Paniagua, Present at American Nuclear Society Space Nuclear Conference, San Diego, California, June 5-9, 2005 [17 pages].

13. Multi-MICE: A Network of Interactive Nuclear Cryoprobes to Explore Ice Sheets on Mars and Europa, George Maise, James Powell, Jesse Powell, John Paniagua and Hans Ludewig, NASA Institute of Advanced Concepts, Phase 1 Report, NIAC Subaward No. 07605-003-047, May 1, 2006 [145 pages].

14. Multi-MICE: Nuclear Powered Mobile Probes to Explore Deep Interiors of the Ice Sheet on Mars and the Jovian Moons, George Maise, James Powell, Jesse Powell, John Paniagua, and Hans Ludewig. Presented at STAIF 2007 Conference, Albuquerque, New Mexico, February 11-15, 2007 [10 pages].

15. MICE: A System of Compact Mobile Nuclear Probes to Explore the Deep Interior of Mars North Polar Cap, George Maise, James Powell, John Paniagua, Jesse Powell, and Hans Ludewig, presented at the 57th International Congress, Valencia, Spain, October 2-5, 2006, paper IAC-06-A3.P3.5.

# Detailed Papers and Reports on MIC

1.  MIC—A Self Deploying Magnetically Inflated Cable System for Large Scale Space Structures, James Powell, George Maise, and John Paniagua. Acta Astronautica 48, No 5-12, pp 331-352, 2001 [21 pages].

2.  Deployment of Large Structures in Space Using the Magnetically Inflated Cable (MIC) System, James Powell, George Maise, John Paniagua, and John Rather, Paper IAC-06-D1.2.09; delivered at the 57th International Astronautical Congress, Valencia, Spain, October 2-5, 2006 [12 pages].

3.  Ibid, viewgraphs of oral presentation at 57th International Astronautical Congress, October 2-5, 2006 [28 pages].

4.  MIC: Magnetically Deployable Structures for Power, Propulsion, Processing, Habitats, and Energy Storage at Manned Lunar Bases, James Powell, George Maise, John Paniagua, and John Rather, to be delivered at STAIF-2007 Conference, Albuquerque, New Mexico, February 11-15, 2007 [9 pages].

5.  Magnetically Inflated Cable (MIC) System for Large Scale Space Structures, James Powell, George Maise, John Paniagua, and John Rather, NIAC (NASA Institute of Advanced Concepts) Phase 1 Report, May 1, 2006 [162 pages].

6.  MIC—Large Scale Magnetically Inflated Cable Structures for Space Power, Propulsion, Communications and Observational Applications, James Powell, George Maise, and John Rather, delivered at SPESIR-2010 International Forum, John Hopkins Applied Physics Laboratory, February 23-26, 2010 (12 pages).

7.  Ibid: Viewgraphs of oral presentation at SPESIF International Forum, February 23-26, 2010 (31 pages).

8.  A Development and Test Program for the Magnetically Inflated Cable (MIC) Large Space Structures System, James Powell, George Maise, and John Rather, delivered at SPESIF-2011 International Forum, Johns Hopkins Applied Physics Laboratory, March 15-17, 2011 (14 pages).

9.  Ibid: Viewgraphs of oral presentation at SPESIF International Forum, March 15-17, 2011 (26 pages)

www.ingramcontent.com/pod-product-compliance
Lightning Source LLC
Chambersburg PA
CBHW070222190526
45169CB00001B/51